THE MAN WHO BENT LIGHT

OTHER LOTUS TITLES

Anil Dharker	*Icons: Men & Women Who Shaped Today's India*
Aitzaz Ahsan	*The Indus Saga: The Making of Pakistan*
Ajay Mansingh	*Firaq Gorakhpuri: The Poet of Pain & Ecstasy*
Alam Srinivas	*Women of Vision: Nine Business Leaders in Conversation*
Amarinder Singh	*The Last Sunset: The Rise & Fall of the Lahore Durbar*
Aruna Roy	*The RTI Story: Power to the People*
Ashis Ray	*Laid to Rest: The Controversy of Subhas Chandra Bose's Death*
Bertil Falk	*Feroze: The Forgotten Gandhi*
Harinder Baweja (Ed.)	*26/11 Mumbai Attacked*
Harinder Baweja	*A Soldier's Diary: Kargil – The Inside Story*
Ian H. Magedera	*Indian Videshinis: European Women in India*
Jenny Housego	*A Woven Life*
Kunal Purandare	*Ramakant Achrekar: A Biography*
Maj. Gen. Ian Cardozo	*Param Vir: Our Heroes in Battle*
Maj. Gen. Ian Cardozo	*The Sinking of INS Khukri: What Happened in 1971*
Madhu Trehan	*Tehelka as Metaphor*
Moin Mir	*Surat: Fall of a Port, Rise of a Prince, Defeat of the East India Company in the House of Commons*
Monisha Rajesh	*Around India in 80 Trains*
Noorul Hasan	*Meena Kumari: The Poet*
Prateep K. Lahiri	*A Tide in the Affairs of Men: A Public Servant Remembers*
Rajika Bhandari	*The Raj on the Move: Story of the Dak Bungalow*
Ralph Russell	*The Famous Ghalib: The Sound of My Moving Pen*
Rahul Bedi	*The Last Word: Obituaries of 100 Indian Who Led Unusual Lives*
R.V. Smith	*Delhi: Unknown Tales of a City*
Salman Akthar	*The Book of Emotions*
Sharmishta Gooptu	*Bengali Cinema: An Other Nation*
Shrabani Basu	*Spy Princess: The Life of Noor Inayat Khan*
Shahrayar Khan	*Bhopal Connections: Vignettes of Royal Rule*
Shantanu Guha Ray	*Mahi: The Story of India's Most Successful Captain*
S. Hussain Zaidi	*Dongri to Dubai*
Thomas Weber	*Going Native: Gandhi's Relationship with Western Women*
Thomas Weber	*Gandhi at First Sight*
Vaibhav Purandare	*Sachin Tendulkar: A Definitive Biography*
Vappala Balachandran	*A Life in Shadow: The Secret Story of ACN Nambiar – A Forgotten Anti-Colonial Warrior*
Vir Sanghvi	*Men of Steel: India's Business Leaders in Candid Conversation*

FORTHCOMING TITLE

Various	*Lockdown Anthology*

THE MAN WHO BENT LIGHT

A MEMOIR

NARINDER SINGH KAPANY

FATHER OF FIBRE OPTICS

Lotus Collection

First published in 2021

The Lotus Collection
An imprint of
Roli Books Pvt. Ltd
M-75, Greater Kailash II Market, New Delhi 110 048
Phone: +91 (011) 40682000
E-mail: info@rolibooks.com
Website: www.rolibooks.com
Also at Bengaluru, Chennai & Mumbai

Layout Design: Sumeet D. Aurora
Production: Lavinia Rao

ISBN: 978-81-952566-0-0

Typeset in Adobe Caslon Pro by Roli Books Pvt. Ltd
Printed at Lustra Print Process Pvt Ltd., New Delhi, India

Acknowledgments

I am deeply indebted to **Tom Parker**, who tirelessly worked for a year towards recording my experiences, authenticated them, and prepared the final draft of my book.

Sonia Dhami, Executive Director of the Sikh Foundation, worked diligently to select the images, consistently reviewed the relevance of the topics, and oversaw this book's production.

Furthermore, I am deeply indebted to **Raj Kapany** and **Kiki Kapany** for steadfastly listening to my experiences and making valuable suggestions.

Kapany Coat of Arms at the family home in Woodside, California

Dedicated to the youth everywhere...

And to my beloved grandchildren,
Ariana Kaur Kapany Schwarz and Misha Kaur Kapany Schwarz
Tara Kaur Kapany and Nikki Singh Kapany

May they learn the teachings of the Sikh prophets and pursue them.

Contents

I. BOYHOOD

Property 4
Sodhi-Wala 7
The Ring 9
The Power of Belief 13
'Here, Bino, Take My Hand…' 15
Stolen Fruit 18
Naughty 22
Almost-True Stories 25
Chain Reaction 30
Father's Radio 32
The Drunken Elephant 34
Deadheading 37
Partition 39

II. LEARNING

Obsessed 52
More Mentors 55
Satinder 58
My Wee Scottish Sojourn 61
Rainy Day Woman 65
Re-obsessed 67
London Snapshots 76
Piled Higher and Deeper 80
The 80-Pound Car 82
The Plan and the Man at the Back of the Room (1) 86
Campari 88

III. COMING TO AMERICA!

Gestation 96
It was a Very Good Year 102
The Plan and the Man at the Back of the Room (2) 104
The Right Thing 113

IV. IN BUSINESS

The Bet 122
Optics Technology 127
Searching for Something New/Veni, Vidi, da Vinci 134
Delhi Again? 140
'If There's a Fork in the Road, Take It' 144
The Little Company that Could… and Did 146

V. TEACHING

'Anything you Want' 154
Making it Happen/Seeds 157

VI. FARMING

A Sikh, a Farmer 172
Wine and Oil 178

VII. BEING SIKH

Sikhism: Equality and Charity 186
The Art of Giving 191
Sikhism: A Spiritual God and an Equal Opportunity 194

VIII. COLLECTING AND SHARING SIKH ARTS

The Gift of Art 198
The Koh-i-noor 201
The Sikh Foundation: Inspire, Educate, Enrich 207

IX. RULE, BRITANNIA!

A Home with a View 216
'I Cooked for You!' 220

X. BELONGING

Belonging 226

XI. MASSACRE

Massacre 236

XII. FAMILY MATTERS

Once-in-a-Lifetime 244
Kids 248
Family Ties 254
Passing 260
Oh, Satinder… 269
Eulogy 272
In Closing 276

PART I

BOYHOOD

Facing page: Watercolour and pen on paper, artist Sumeet D. Aurora

I was born Narinder Singh Kapany to Kundan Kaur and Sundar Singh in the town of Moga, in the province of Punjab, on the subcontinent of India, on 31 October 1927.

I am now ninety years old. I have written this book to capture some memorable moments and scenes from my life. But first, a few words about memory. When prompted, as I am frequently, by my son, Raj, or daughter, Kiki, I recall details from days long gone by, sometimes fine, sometimes coarse, sometimes, without being aware of it, details that may be more imagined than actually remembered. But always, always they are details in search of the truth.

I offer, for example, those memories evoked by my passage with my wife, Satinder, from England to New York on a German luxury liner. Memories of that cruise and of the provenance of the vessel – as provided by one of my fellow passengers – live as clearly and vividly in my mind as do those of my most recent dinner with one of my grandchildren. And yet, the entire voyage, at least as I initially remembered it, could never have happened. The Fuhrer's aviso that I recall sailing the Atlantic on was decommissioned and dry-docked in 1951, and Satinder's and my crossing took place, without doubt, in 1955, following our marriage. Confronted with this fact, I thought perhaps there was a second German liner that was twin to the first. No such finding in web searches. And yet, the journey itself is still so indelible, the writing on the page – before I modified it in the service of the truth – so very convincing.

And then I realized: but, of course! These are *my* memories. If they are faulty in places, so be it. So what if I conflated a story I'd once heard about a German dispatch boat with that of my own crossing of the Atlantic on a larger Cunard liner?

What matters is that I am able to capture the spirit of the person (or the ship) and his (or her) essential truth, and the role that he played in my life, and the place he takes up in my occasionally flawed but, I'd like to think, rich, powerful ability to recall the past.

And so I begin…

Property

Shhhh! Quiet. If you listen carefully you'll hear them. My grandmother – my manji – and my Auntie Nuntal. Up on the roof. Four floors up.

It's 1930, or thereabouts. I am roughly three years old. I am at my mother's family home in Moga, our 'ancestral home,' she called it. I know how to walk – I'm *three*, after all. Still, my memory has me *crawling* up the carpeted stairs toward the voices in the late-afternoon light that is streaming through the open door on the landing leading to the roof. The further up I go, the brighter the light.

It is summertime in Punjab, in North India, and the heat is unremitting. Everyone else in the grey, four-story, brick house – my mother Kundan, my brother nicknamed Jat, my Auntie Vido, and the servants – is asleep, passed out from the heat. My father Sundar has not made the journey with us from our family home. The temperature is more than 100° Fahrenheit, according to the new thermometer in the parlour that Father taught me to read.

At the open door to the roof, I pause and see Grandma – my manji and my Auntie Nuntal, standing at the far edge of the roof. They are both gesturing to a spot on the horizon. Not far from where they stand is a large, ancient, wooden table that once served as the main table in our first-floor dining room. A half-dozen, black rattan chairs are scattered higgledy-piggledy around it. A faded pink umbrella, once bright red – or so Mother told me it had been, back when she was a young girl – planted in a concrete base, casts a meagre shadow on the table, while a single, small, near-leafless tree in a half-barrel of dry dirt points feebly to the sky.

Standing now on the threshold, I can clearly hear what they are saying, only I can't make any sense of it – particularly a single, unfamiliar word spoken again and again: 'property.' Embroiled as they are in the heat of the day and in their discussion, neither woman notices me until I am standing right next to them.

It is Auntie Nuntal who spots me first. As if I weigh nothing at all, she scoops me up in her arms in one swift motion. She is a tall woman, and I have a sudden sense of being at a significant height. She is the second favourite of my two aunties, but I love them both, truly. Auntie Nuntal is the younger one, about seventeen, more talkative, more argumentative, more sure of herself. I will attend her wedding years later.

'My goodness, what are you doing up here, on the roof by yourself?' Auntie questions. 'Don't you know how dangerous it is?' She smells like spice.

My manji, a heavyset woman both generous and loving, strokes my cheek as Auntie Nuntal clutches me tighter to her bosom, nearly suffocating me in the heat. 'Well, he's not alone now,' Manji reassures, cooing, 'my little Bino' – my family's pet name for me.

From my elevated perch, I have no sense of danger whatsoever. A substantial masonry barrier encircles the entire perimeter of the roof. Other than the door I just passed through, only a single, latched gate along the barrier offers an alternate way down, and it is secured by a rusted padlock. A large, black bird flies overhead, squawking as if lamenting the heat and his dry-feathered lot in the grand scheme of things.

'What is "property"?' I ask. 'Land,' Auntie Nuntal offers.

'Land*s*,' Manji corrects. 'Auntie Nuntal and I were talking about our family's properties that were given to us generations ago by a powerful and wealthy Sikh maharaja.' She pauses for me to register the import of such a magnificent bestowal. 'Someday, they will be yours.'

As it turned out, this would not be the case. Only a generation later, my two uncles, Gurbaksh and Balbir, both of whom I remember primarily as being tall, were careless with the family fortune they inherited and eventually sold off much of the land to pay their debts. But that was still coming some years later.

'Show me!' I insist, as Auntie Nuntal sets me down. 'Show me the properties.' Now Manji stands behind me, her hands on my shoulders, turning me around in a full circle and gesturing to the horizon in every direction. 'There!' she says, 'and *there* and *there* and *there!* As far as you can see.' I follow her gestures, registering a vast and seemingly endless quilt of variegated green fields and a few outcroppings of small structures all the way to the horizon in every direction. 'Everything you can see, my Bino, belongs to us, belongs to our family.'

I stand in awe. Can this be true? How can one family possess so much? And what does that mean for me and for my future life? I have no answers to these questions, of course. Since that day I have realized that I had, in fact, learned something profound under that softly darkening summer sky: that mine was a life full of possibility. And, but for the curvature of the earth, my horizon would have been even more distant. But as dusk turned to night and it became darker and more and more difficult to see even a few feet in front of me on that roof, I'd also worried about how easily and how quickly all these properties could be lost.

Sodhi-Wala

The next morning, Mother rouses me from my sleep. It is still dark. My older brother, Jat, is already awake. I can see him behind Mother, making silly faces. Mother switches on the light. 'Quickly, Bino. Quickly to breakfast,' she says. 'Today we are going on a journey.'

'To where?' I ask sleepily. 'To Sodhi-Wala,' she says.

'But isn't that Manji's name? Sodhi?'

'Yes,' she says, 'but it is also the name of a nearby village. One of our properties,' she explains. 'Now hurry.'

Presently, a servant brings an open, wooden, four-wheeled wagon to the courtyard in front of the house. It is drawn by two massive, double-yoked white bullocks. The driver sits up front, and in the rear my mother and my two aunties, Vido and Nuntal. All three sisters are dressed in similar salwar kurtas. Each carries a striped parasol.

Clambering aboard dressed in shorts and short-sleeve shirts, Jat and I take our places on the floor of the wagon. A large picnic hamper partially covered in a patterned orange batik cloth sits directly behind the driver. At first, as the wagon begins to roll, I am quite content to stand, trying to keep my balance as the old wagon with its wooden wheels creaks down the well-worn dirt path. Jat does the same, as the wagon rolls past an ancient British car abandoned at the side of the house, grass and weeds growing through the floorboards, rust intermingling with the remains of its flaking black paint.

Deftly, Jat and I keep each other from falling in the moving wagon until one wheel careens off a large rock and we both tumble to the wagon floor, where, instead of continuing our entertainment, we soon fall into a half-sleep, soothed by the chatter from the three sisters above. As the sun rises higher in the sky, the chatter turned to murmurs until there is hardly a sound to be heard other than the creak of the wooden wheels and the pitching of the wagon.

The villagers of Sodhi-Wala greet us in an open, freshly mowed, grassy field with a bright green water pump at its centre, where we stop to lunch and to water the bullocks. Though the village of bears our name – or we bear its – I soon discover that, though the villagers speak the same language as Jat and I do, they are almost impossible to understand.

Of greater interest to me and Jat both is a small shrine with a large polished stone plaque, bearing the Sodhi name and dedicated to my mother's ancestors. It stands not far from the centre of the village, under the shade of a large tree. Auntie Vido explains that on holy days, the men from Sodhi-Wala and other villages nearby would drink and dance on the flat surfaces of the memorial. 'Sikhs are cremated when they die,' Mother adds, 'leaving those who survive the dead only ashes and memorials like these.' I work to understand what she'd said when Jat boasts that he already knows. '"Cremated" means burned,' he explains.

The Ring

Father...
Long before I was born, in the First World War, my father was in the Royal Air Force fighting for the Allies against the Germans. It's how I came to have this ring, one I've worn on the second finger of my left hand nearly every day for the past seventy years. Father wasn't a pilot, however, but a photographer, relegated to the tiny rear cockpit of the dual-wing trainers they flew back then, the forerunners to the famed Gypsy Moths of my boyhood. By contemporary standards, these tin lizzies of the air were rather quaint, buzzing through the skies, dropping small incendiary bombs and dogfighting with their German biplane counterparts – the Red Baron, and all that.

Flying reconnaissance, my father, with his huge box camera, returned to the base having filmed any military installations they might have encountered or an occasional flotilla of German troop supply ships on the open seas. After he developed this film and shared it with his command, a thorough air attack ensued, followed by another reconnaissance flight, with my father onboard to confirm the damage. That was air warfare, a century ago.

Ah, but what about the *ring*?

The story goes that one afternoon, flying over enemy territory (what country, I don't remember; along the Russian front, though), my father's plane crash-landed in a field after being hit by German antiaircraft fire. It was nearing dark before my father and the pilot managed to free themselves from the aircraft's shattered superstructure – largely wood with some steel cross-members. All the while bullets were streaking overhead, barely clearing the tall grasses that were their only shield against the enemy.

As per standing order, the two men, crawling on their bellies, set fire to what was left of the plane to keep the enemy from making either technological or propaganda use of the aircraft. By the time its gas tank exploded, the two men had already crawled a few hundred yards from the wreckage, while bullets still whizzed by in volleys, an occasional tracer lighting the way. Then, for no apparent reason, the bullets stopped. Bad instinct and curiosity – or so Father told it – encouraged him to raise his head to reconnoitre the scene. In that same instant, the pilot rose up on his elbows to pull my father's head down when, midsentence – 'Put your goddamn head back…' – a bullet likely meant for my father hit the pilot in the forehead, scattering blood and the soft mush of his brains into the dry grass around them.

After that, there was no more firing, and my father, afraid, bereft, and exhausted, lay in the grass until it was pitch black. By then, the wreckage lay smouldering – nothing to be reclaimed, the Bosch must have decided, so they abandoned the scene, leaving my father alone somewhere along the Russian front.

Hours passed before he made a move. But then, spying a single light flickering a few miles distant, he decided to make his way toward it, hoping it belonged to an Allied civilian, though someone not as lost or as hungry as he. Besides, what were his choices?

Presently, the clouds cleared and the starlight revealed a rutted wagon path leading to the flickering light, which, as it turned out, stood in the window of a small hovel where an old woman sat quietly at a crude table, the cabin's sole piece of furniture other than a chair and a straw bed. Father's hunger managed to translate his need for food into Russian. Her need for someone to help her find a place safe from the German marauders managed to translate itself into English, which he could understand.

And so it was that the two set out under the cover of the three hours of darkness that remained, heading for an Allied camp, my father helping her carry the few bags of her possessions and using the stars as his compass. Upon arriving at the camp, she produced a ring from a chain that she wore around her neck, beneath her tunic, and an amulet from her pocket, giving both to my father as thanks for leading her to safety. 'For luck,' she told him, as translated by a Russian quartered at the camp. The amulet has long since disappeared – fallen from a pocket, left in the corner of some drawer, loaned to an unappreciative child, who knows?

As for the ring, upon returning to India at the end of the war, Father had given it to his mother, my grandmother, claiming that it would bring her luck. But after she'd worn it a few months, she decided it was too big for her and placed it in a jewellery box. There it remained for a number of years, before she passed it on to her daughter-in-law, my mother, telling her the story of the lucky ring's provenance. Lucky or not, the ring found its way into my mother's jewellery box, where it remained until the late 1940s, when I was in my early twenties and in need of some good luck myself.

Back then, before you were awarded your college degree, you had to pass an exam. If you succeeded, your assigned roll number was published in the local newspaper. If you failed, your number did not appear. By my last year in college, I was fairly certain of my abilities and was surprised and crestfallen when my roll number wasn't among those listed. When I told my mother that my number hadn't appeared, and that I had likely failed the exam, she marched me to the jewellery box, handed me the ring, and without much by way of explanation, said, 'This was your father's… his lucky ring from the war, a reward from an old woman for leading her to safety in the middle of the night.'

So without much to do, other than to suggest that this 'lucky ring' mumbo-jumbo was not the sort of thing that young, sophisticated Sikh men typically believed in, I put the ring on, only to open the newspaper the next day and find my roll number at the top of a revised list, along with a printed apology to those of us whose numbers were left out the prior day.

Over the years, I've been tempted to take the ring off many times – too big, too tight, too flashy, too plain – but for my promise to my mother, who passed away at eighty in 1984, that I'd wear the ring my life long. As for it bringing me luck? So far, so good.

Narinder S. Kapany wearing the ring.

Narinder's father, Sundar Singh Kapany, as a young officer in the Royal Air Force during World War I.

The Power of Belief

...and Mother...

While Father may have been my boyhood hero, Mother was, quite simply, my mother. Much adored, deeply appreciated, always there and, not at all incidentally, loved, she was a Sikh woman of her times. There were six of us children, plus my father, and she loved us all equally, and equally well, as far as any of us could tell. She would also do anything in her power to make us the best that we could be. And we all knew it. We all derived strength from her. Were it not for her and her belief in me, I would likely have languished in some government post in India instead of blossoming into an entrepreneur in the USA.

That said, I remember Mother best from the time I was about five years old and my brother Jat and I were suffering mightily from a fever that she feared would take our lives. Distraught beyond measure, she picked me up and carried me to the large room where the painting of our Guru, Guru Gobind Singh, hung, and before his eyes she promised that were Jat and I to be cured, she would take the two of us along with my father on a pilgrimage to the Golden Temple. Instead of sharing her intense gaze at the Guru, I remember looking at her. I felt her warm breath in my ear, her firm fingers on my rib cage, and knew on some level that I would be saved – and that it would be as much my mother's love that would make it possible as whatever power the venerable Guru might be able to summon.

Narinder's mother, Kundan Kaur Kapany with other members of the family.

'Here, Bino, Take My Hand...'

As a child of six or seven, I was awestruck by the sheer size and amazing construction of the Golden Temple in Amritsar. Today, more than eight decades later, I still am, and a giant painting of the sprawling temple complex – with its central shrine surrounded by a vast walled lake, each side the length of two football fields – hangs behind my desk. The painting is as wide as the desk itself, the temple's perimeter walls appearing to frame me. Seated at my desk, I am *of* the temple. And the temple, it seems, is a part of me.

I have visited the Golden Temple at least twenty times in my life, and with each visit come new memories: the intricate black patterns inlaid into the walkways; the dazzle of the temple's gilt upper surfaces offset by the mellow creaminess of marble; the scent of flowers and spices, of saffron in particular, emanating from the nearby kitchens that welcome all travellers; the soft, insistent murmur of prayer; the graceful arches that serve as portals to the peaceful world within and, as seen from within, to the roiling, more dangerous world without; the golden light above the sarovar, or Pool of Nectar and, reflected in it, the pale, pale blue of the sky above. And, of course, the memory of the Sikh massacre that took place there in 1984. But that comes much later.

Of one particular trip to the Golden Temple taken with my family around 1933, my memory is not of the majestic shrine or the soft-spoken pilgrims, but of Father and my older brother, Jat, and me. Beyond that, I recall little but unremitting sunshine, soft, white puffs of clouds, and a storybook blue sky: a child's drawing.

It is a hot day. Mother sits on a low bench with my sisters some twenty yards distant. 'Go with your father and Jat,' Mother urges me, 'go into the water. Be your father's boy.' Of course, I dearly want to be my father's boy, but I am frightened. Not *that* frightened, however, as it would soon turn out.

Men back then had no special bathing outfits, and would simply strip down to their underpants to get into the water. On this day, I remember Father sitting on the marble coping at the edge of the sarovar. There were steps to lead you into the water gradually, but from what I could tell most of the pilgrims just lowered themselves into the deep water directly from the edge. Neither Jat nor I had ever been in the sarovar before. Sweat pours down my forehead and into my eyes, creating first a tear through which I cannot see, and then a kaleidoscopic dazzle as the sun is refracted by my tears, and everything I see appears fragmented, kaleidoscopic, including Father, who slowly lowers himself into the water until he is still able to stand on the sarovar's bottom, but barely. His chest is broad. His arms look strong. He looks like a hero in a movie.

'Come,' Father says, holding his arms out to Jat, who is standing next to me. Two years older than I, Jat seems eager to take Father up on his offer. Father bends toward him. I watch carefully, almost as if in a dream, still not quite capable of seeing clearly as Jat makes his way to the sarovar's edge. We are only a few feet from each other. The sun must be as hot on him as it is on me. A dip in the cooler water will surely be refreshing. And yet, just as he is about to take Father's hand, Jat withdraws his own. 'Come,' Father repeats, holding his open hand out ever closer to Jat. But instead of taking it, Jat looks back to where Mother is sitting. She shades her eyes to see him, then gestures him back to her. He runs all the way.

'So, *you* then,' Father addresses me. 'Here, Bino, take my hand.' He glances over at Mother. 'Come, won't you join me?' This time when he holds out his hand, it is to me. Still, I look around me to see if there are any others. But no, Father's hand is mine for the taking. So I take it, thinking, 'Why not? What can go wrong?' And in the next instant, he pulls me up and over his head, and *splash!* places me flat into the water on my belly, his two arms now cradling me as I paddle like the other children I had been studying for what to do next.

'Me, now,' I hear Jat offer from the edge. 'Me next!' But it would be a while before I relinquished my turn, lying on my back now, looking up into Father's eyes.

'The Holy Temple,' in *Original Sketches in the Punjaub by a Lady*, 1854, Colour-tinted lithograph, 36 × 27 cm, Kapany Collection.

'Entrance to the Holy Temple at Umritsar, from the Gate of the Kutwallee,' in *Original Sketches in the Punjaub by a Lady*, 1854, Colour-tinted lithograph, 36 × 27 cm, Kapany Collection.

Stolen Fruit

Years after I discovered my auntie and my grandmother on the roof of their ancestral home talking about property, I was visiting my cousin Jagdish in my paternal grandfather's home, in what was then the Indian State of Patiala. Grandfather was a sessions judge in those days – a learned jurist who would write a number of books. Back then India was not yet independent but still British, and maharajas were the major landowners and de facto rulers, with those in the part of India where I was visiting being primarily Sikh.

Both of us about eight years old, Jagdish and I quickly grew bored with parlour games and playing hide-and-seek in the palatial home where he lived. In the warm, often insufferably muggy evenings, we shared the same room, each of us lying on our back, watching the blades of a large ceiling fan stir up whatever breeze it could. It was a tedious thing, let me tell you. So when Jagdish suggested we get up with the sun the following morning to steal some fruit from the maharaja's orange groves a few miles down the road from Grandfather's house – the largest and most perfectly paved road in the region, surrounded on either side by neatly manicured trees hiding the extensive fruit groves – I was more than game.

Nearly at our destination that next morning, Jagdish and I suddenly heard coming from behind us the hoofbeats and snorts of horses and the leather squeak and metal rattle of harnesses. We quickly made way for them to pass, a procession of ten beautiful, ornate, horse-drawn buggies making their way down the road. At the lead was the most elegant, with a driver and two passengers, the first a Sikh maharaja – Bhupinder Singh of Patiala, I would later learn – who sat perched on a golden cushion, and next to him a second Sikh, the two men deep in conversation. It was the turban that marked them as Sikhs, or so I thought. As the procession passed us by, I noticed that virtually everyone was wearing a turban, including the passengers in a considerably lesser buggy, wearing Western clothes.

'Quick!' Jagdish called to me, grabbing my shirt sleeve and urging me behind the closest tree. I'm sure if someone were actually looking for us, they would've seen us immediately, two young boys trying to hide behind the slender trunk of what was, after all, only a decorative tree. But no one in this stately parade was looking for the likes of us. And in less than a minute the procession had passed, looking to me like a mirage, as the last buggy disappeared into its own dust. I couldn't quite believe it. First, there was the procession itself: so many buggies, so stylishly outfitted, so early in the morning, and passing just a few feet away. Second, there was the taste of the oranges: warm, sun-blessed, incredibly sweet. Third, that no one, it seemed, but Bhupinder Singh himself, faced anywhere but forward, nor said a word to another. And fourth, that this entire procession was all Sikh. I'd never thought one way or another that being a Sikh was important, yet here was this august group, looking all the while like kings of the universe – or at least royalty of some sort.

Much like the property my grandmother pointed out to me on that rooftop years earlier, these distinguished Sikhs aroused a sense of possibility in me. Perhaps I, too, might someday be able to ride in such a distinguished procession.

At the dinner table that evening, the first thing that my uncle Mangal Singh asked Jagdish and me when we told him what we'd seen that morning was what we were doing on the road. And where had we been for breakfast that morning? Sheepishly, Jagdish volunteered that we'd gone to sample some fresh fruit from the maharaja's overly abundant orange trees; Jagdish seeming to convey that we'd done the maharaja a favour that morning by 'pruning' his grove.

In any event, I didn't want our petty crime to outshine what I thought to be an important discovery. 'But they were all Sikhs,' I insisted. 'The maharaja and all the important men. Sikhs.'

It was Uncle who set me straight. He told me that the maharaja was indeed a Sikh, most likely Bhupinder Singh. And that upwards of a half-dozen maharajas in North India were also Sikh. But the others in the procession, possibly local secretaries of this or that, were likely only wearing Sikh attire in deference to the maharaja, particularly the British-looking, Western-dressed ones.

Still, I was not to be disappointed. At the age of eight, I'd seen a procession that looked royal and regal, with passengers who wore the clothes and the turban of my own family. It is an image that remains with me to this very day: for the impression it made so very, very long ago, and the fact that Bhupinder Singh's grandson, Amarinder Singh, is now the Chief Minister of Punjab.

Left: Judge Nagina Singh, paternal grandfather of Narinder.
Right: The Law of Confessions, by Sardar Nagina Singh.
Photo courtesy: Deepinder Singh Kapany.

Narinder's great grandparents with his grandfather (middle).
Photo courtesy: Deepinder Singh Kapany.

Naughty

I'm not sure if I made it clear. My cousin Jagdish – the one whom I accompanied on the fruit-stealing adventure when we were both around eight years old – was a very naughty boy, much naughtier than I, always up to one prank or another. One day, that same year we stole the maharaja's oranges, Jagdish came into possession of an enormous bar of chocolate wrapped in foil and covered with a brown paper sleeve. Only it wasn't really chocolate. It was a chocolate-flavoured laxative that Jagdish's auntie would regularly nibble on to help encourage her own widely discussed gastrointestinal activity.

One person in the household who definitely had little interest and, most likely, no knowledge of Auntie's digestive goings-on was her brother-in-law Uncle Hari Chacha, our family's 'crazy uncle.' You know, the fellow who drinks too much, who eats everything, who seems unable to hold a job, who always says the wrong thing, who speaks too loudly and too much, and who smooches all the women, thinking he is quite the ladies' man.

On this, the last night of my family's visit to Patiala, we were sitting around the parlour when Jagdish sidled up to me and said, 'Come, Bino, I have a plan.'

'Better than stolen fruit?' I queried. I could still taste the contraband oranges from the maharaja's grove, still picture the somewhat-Sikh procession.

'*Much* better.' By now, Jagdish had revealed the laxative bar and removed the outer paper sleeve from it, leaving only the tinfoil. He then waggles the brightly waggled treat in the direction of Uncle Hari, who was surreptitiously gulping whiskey from a flask in a dark corner of the parlour.

'For me?' The glint from the foil caught his greedy eye. 'What?' he asked, 'chocolate?' Jagdish nodded, then broke off a small piece, as if to share it with his uncle, knowing that, given the opportunity, Uncle would leave Jagdish with a small square and snatch the remainder, eating virtually the entire bar, and make short work of it. True to form, that is precisely what Uncle did, devouring about five days' doses of laxative in a few seconds. 'See, Bino,' Jagdish said to me, chuckling, 'I told you it would be better than stealing a few oranges.'

Satisfied with his postprandial snack, Uncle Hari stood and made his way to his bedroom, leaving a fulsome burp in his wake as Jagdish and I exchanged glances.

'There is more to come,' my naughty cousin insisted, producing the key from his pocket and quickly locking Uncle Hari's room from the outside.

'But what happens if he has to…' I stopped midsentence, knowing the answer.

'…to use the toilet?' Jagdish offered. He smiled the smile of the truly naughty.

It is not until Jagdish and I have waited in the vicinity of the door for a few hours that he decided to accelerate the action. Grabbing the family's cute, though highly spirited, pet Rhesus monkey from its nearby cage, he instructed me to unlock and open Uncle Hari's door – which I did, but only a crack – and he threw the monkey in. Freed from his enclosure, the monkey quickly laid bare every surface and shelf in the room, books and other stuff scattered willy-nilly while Uncle Hari rattled the doorknob frantically to be released.

'Let him out,' I finally insisted after we had each gotten beyond our laughter. But I am alone in my sentiment – the less naughty boy, as it turns out. 'Let him out,' I insisted. 'Let him go to the toilet.' And I did. But only over Jagdish's objections.

The two of us, Jagdish and I, remained the closest of friends our entire lives, until he died a few years back, in Great Britain.

Family pet monkey 'Budhoo' and cat 'Kali'.

Almost-True Stories

As a boy of ten, I would often listen to my manji – in this instance, my father's mother – reading aloud from one of two antique volumes in the parlour of our ancestral home to an assemblage of women from the local area. Manji's voice was melodic and full of life, as were the stories she read, passed down over the centuries through my family. These volumes, called Janam Sakhis, are a collection of stories that chronicle the life of Guru Nanak, the founder of the Sikh religion. They were commissioned, some 200 years ago, by my ancestor who was the head of the Takhat Patna Sahib, a major gurudwara in the city of Patna.

Not only were the stories contained in these two books animated and colourful in their tone and descriptions, they were also brilliantly illustrated, each volume containing forty magnificent paintings. A skilled storyteller in her own right, Manji carefully passed the volume from woman to woman, urging each to touch the pages to commune with Guru Nanak's life.

'If you touch the pages now, you will be sanctified,' she promised. This led to a collective swoon among the women, whose prior exposure to the great Sikh Gurus was minimal, if that.

It was often at this point, when the parlour was in the collective thrall of Grandmother's reading, that my father would enter the room and brusquely announce, 'Mother, what you are telling your guests is *not true*!' – often to the shock of those assembled – then disappear as quickly as he had appeared, leaving Grandmother to pooh-pooh Father's claims and try to recreate the earlier sense of rapture in the parlour by reading yet another story about the great Guru Nanak.

Later, after the women had left, Manji would chide Father for interrupting, but he was always unrepentant. 'What you are reading to these women is not true,' he insisted. 'The stories are wrong!' At his core, Father was more a man of science than a man of the spirit. Which is to say, he believed in the Guru's wisdom, just not the fantastical stories that accompanied it and that, he felt, had been fabricated to further deepen the belief of Guru Nanak's followers.

Whether the stories were 'true' or not, I felt concerned about the apparent disagreement in our family. 'See here, Bino,' Father said when I asked him who was right about the truth of the stories, he or Manji. 'See here,' he said, opening the first volume to lend credence to his frustration. 'See, in this story of the Guru Nanak, the one where he arrives at the holy city of Haridwar during an unendurable heat wave and comes upon some priests splashing water from the River Ganga upwards, in the direction of a relentless sun. "What are you doing?" the Guru asks. "Sending water to cool the sun," he is told. The Guru considers this and then, cupping his hands in the water, he splashes it sideways. When asked by the priests what *he* was doing, he said, "If you can send the water to cool the sun millions of miles away, I can do the same with the water to irrigate my distant parched fields."'

I had been listening carefully to Father's story of the Guru, but now I interrupted. 'Yes! But all that *could* have happened. What's wrong about it?' I was concerned that I was not understanding.

'True,' Father agreed. 'But the story, as written, does not stop there. It continues. It goes on to say that the water the Guru tossed in the direction of his faraway fields actually *did* land there, and that his plantings were saved. And *that*,' Father emphasized, 'is the part that was wrong. That's where a story, *possibly* true, became *impossible*. A *miracle*… and,' he concluded, 'not to be believed.'

'But that is just one of the stories in the book, Father! Manji says there are many more in Patna's two volumes.'

'All with like endings,' Father insists, '*possibly* true up to a point, and then pure nonsense. A miracle! Posh!'

And then he tells another. 'There's the one about Guru Nanak walking with a disciple to Mecca. Exhausted from the journey, they lie down to rest in the courtyard of the mosque, whereupon they are taken to task by a priest for lying with their feet pointed towards the Kaaba, the room in the middle of the mosque known as the "house of God." Guru Nanak apologizes, and asks the priest to move him so his feet point to the side of the room where there is no God.'

I interrupt again. 'Yes!' I insist, more certain of myself than when I interrupted the prior story. 'All that makes sense.' And, perhaps precociously for a ten-year-old, I continue, 'And it teaches a lesson: that God is everywhere.'

'Ah,' Father says, clearly pleased that he is about to be vindicated, 'but the story in the book goes on to say that every time the Guru moves his foot in another direction the Kaaba *itself* moves along with it. Again, a miracle. An untruth,' he insists, 'reached by people who didn't really understand where Guru Nanak's mind was.'

Back then, and over the years since, the more I thought about it, the more I came to agree with Father about Sikhism, and about the other great religions of the world. Truths are revealed, object lessons are taught, and the magic is exposed. As for the miracles – men and women having actual discussions with God, a Guru splashing water from a fountain to irrigate his distant fields, a foot repositioned and with it the entire city of Mecca – I concluded they are there simply to make them better stories.

Meanwhile, despite Father's protestations about her book clubs, Manji paid him little heed. Rather, she rescheduled her readings to the village women for times when she knew he'd be out of the house. Years later, when she was a very old woman, she entrusted the care for the two volumes to her son, my father. And about forty years ago, when I was fifty – older than Father was when I first learned about Guru Nanak – he gave the two splendidly illuminated volumes to me.

Today, one of those volumes, with its magnificent illustrations, sits atop a table in my home, its art near-priceless, its stories... well, *possibly* exaggerated. And its companion volume resides in San Francisco's Asian Art Museum, to which I donated it some fifteen years ago.

Janam Sakhi manuscript, India, *ca.* 1800–1850, Kapany Collection.

Illustrated Janam Sakhi, carbon ink on paper with gouche, gold paint, leather binding, *ca.* mid-nineteenth century, 19.1 x 13 x 6.4 cm.

Guru Nanak's wedding, folios from a Janam Sakhi manuscript, Lahore, Pakistan *ca.* 1800–1900, opaque watercolours and gold on paper, 20.3 x 16.5 cm, Asian Art Museum of San Francisco, Gift of the Kapany Collection, 1998.58.8.

Chain Reaction

I blame my penchant for pranks on my cousin Jagdish. I actually used to try to out-naughty him. In high school, in particular, I often played pranks on my teachers, but there was always my history teacher Mr. Zachariah to help clear my name. No matter how foolish I made my teachers look, no matter how foolish I made *myself* look, Mr. Zachariah was there to the rescue, making things right with the other teachers and even with the principal.

One day he even helped me keep the lead in the school bicycle race around the track that encircled the soccer field. I was a fast bicyclist and was ahead of the others – at least a full lap – when my chain broke and my bike slowed to a halt. That's when I heard the call, 'Narinder, Narinder!' It was Mr. Zachariah, running toward me with his own bike. 'Here,' he said, handing his bike to me and grabbing my immobilized one, 'take mine!'

'Are you sure?' I couldn't believe my good fortune. Meanwhile, a phalanx of my competitors pedalled frantically by.

'Here!' he commanded. 'Go!' And just like that I was back in the race, in moments passing everyone in the pack who had shot past me, taking the lead once again. Mine was a short-lived victory, however. Not a victory at all, actually, when, three quarters through the race, the chain on Mr. Zachariah's bike also broke, and I had to retire from the race, once and for all.

Years later, when I was already living in America, on a holiday back home in Dehradun, I invited Mr. Zachariah to my family home. By that time our age difference of approximately ten years didn't seem so sizable, and I felt that Mr. Zachariah was almost my peer, though he was now the school principal and a somewhat more formidable figure merely by dint of the title. Still, it didn't keep me from confessing. 'I did some very naughty things in school; and in your class, you know.'

'Of course, I know, Narinder,' he said. 'And you know what?'

I truly didn't, 'know what.' So I asked, 'What?'

'I liked it! You had such great spirit. And I knew that someday you would do something very unique. And,' he paused, 'look!'

I did. I looked at him, at his open, welcoming expression, his wide smile, and his evenly spaced white teeth.

'And look!' He repeated. 'You *did* it!'

'And you were my *mentor*,' I replied.

'Call it what you like. I just had faith in you.'

Narinder in his teens.

Father's Radio

Father, again.

I'm now nearing ninety-three and still there is rarely a day I don't at least have a fleeting memory of my parents, Father in particular. He was a great man and, more colloquially, a wonderful guy. Physically and emotionally strong. Loving. Strict. He administered the occasional spanking, but we, all of us children, could tell his heart wasn't really in it. More true to who he really was, was the man who would embrace us and embrace our mother, so we could all see that it was intimacy that he favoured. Ours was a loving family, at once traditional and modern. None of us ever hungered for affection.

After leaving the British Air Force, Father became a contractor. When I was a little kid I didn't precisely know what that meant but was content to imagine it: something about building, about forging something from nothing. I liked that Father had those skills and hoped that I did as well.

What I most remember of Father during those preteen days before I would head off on my own adventures was the sound of his motorbike: Father making his way back home, me waiting for him in the soft, muted evening, the quiet broken by the gravelly gargle of the bike's exhaust. I imagined I could hear him miles away, though in fact I probably couldn't.

Father also loved walking, often with a walking stick, either by himself or otherwise accompanied by one or more of his friends. Years ahead of his time, he took these constitutionals for the sake of his health. Many of my memories of him are as he is just setting out on a walk, a spring in his step. Either that, or his arrival back in the courtyard, like a distance runner crossing the finish line, thankful, triumphant, tired.

Father also liked simple rituals, practicing one virtually without fail: closing himself in the parlour before dinner for at least a half hour to listen to the news on the radio. Conversant as he was in the events of the day, Father gave the impression of being a learned man. And he was, though I was well into my teens before I discovered that my father's primary reason for locking himself away in that room was to have a whiskey – or two, or even three – if I'm to believe my mother. Drinks or not, he always comported himself elegantly. I don't think I ever saw him drunk.

Sundar Singh Kapany as a young man.

Sundar Singh Kapany.

The Drunken Elephant

I may have never seen Father drunk, but I knew what 'drunk' meant. I'd seen others inebriated, emboldened by various brews. One particularly memorable time was when I was with my cousin Kulwant Singh. We were both fifteen and I was still living in Dehradun. It was evening, the darkening sky tinged with red, and we were walking down the road, when off in the distance we spotted an elephant. It was heading our way, lumbering cantankerously, swaying from side to side, and being ridden by a bare-chested mahout, sitting tall in the saddle, or the howdah as it is known.

He had obviously seen us as well, because he headed right for us, drawing closer and closer until man and elephant were only a few yards away. Stopped and sitting sideways facing us, the mahout dug his spurs into the elephant's haunches. The beast groused, then lifted his trunk high into the air and trumpeted.

'Well,' the mahout said, 'what are you boys waiting for? An invitation from the Viceroy?'

Again, the driver goaded the elephant, though this time he didn't grouse but knelt before us. Acknowledging the driver's request, I asked, 'Are you free?' I don't really know why, but I dearly wanted to mount the pachyderm. And it was clear that the mahout knew it.

'Nothing is free,' the man countered. 'Just two rupees.' He held his hand out to us – 'each.'

The transaction complete, Kulwant Singh and I mounted the elephant, he behind the mahout in the saddle, and me in front of him. We had barely gained our bearings before the mahout goaded the elephant again with his spiked spurs, though this time instead of kneeling, the elephant spun, did a clumsy dance, and turned back in the direction he came from, again lumbering and rocking none too surely, it seemed, on his feet.

At last fully aboard, I was now finding it hard to hold on. I looked over my shoulder to see my cousin. He appeared worried. Still, the mahout persisted in goading the elephant, on and on, again and again. 'What's wrong with you?' I finally implored. It was then that I smelled the liquor on his breath.

'*Wrong?*' The man seemed confused, then not. 'Just some ale at the festival. And then some more.' That instant, and without warning, the elephant careened off to the left, quickly righting himself for a few moments, and then, seriously listing, veered off to the right.

'He's drunk, too!' the man slurred, then laughed, a sloppy, deep, guttural guffaw. 'He's had more to drink than me,' the man confessed, as the elephant lurched again. I grabbed the howdah to avoid being pitched off.

'Let us off!' I demanded. And as the mahout slowly, drunkenly registered the request, I turned to Kulwant and yelled, 'Jump!' as he disappeared from sight.

I did the same, miraculously landing on my feet. Kulwant, meanwhile, was already running back to where we'd first mounted the elephant, but for some reason I stood transfixed. Just then, the elephant reared up on his two hind legs only a few feet away from me, exposing his huge underbelly and trumpeting loud enough to hurt my ears.

It was an image and a sound that I was not soon to forget: that massive, drunken beast with his besotted driver, close enough with his raised forelegs to pummel me into the ground.

It was then – and it remains now – one of the most fearful moments in my life. For months I would dream it, culminating with elephant and driver making their way hastily into the night until they were more mirage than flesh and blood, and the whole event seemed more imagined than actual. It wasn't until months later, on the advice of a friend, that I had finally told the story enough times to literally get it out of my system.

Until today, that is, when that mouldy, grassy reek of the animal and the demented, fermented laugh of his driver come once again to life and the familiar fear returns.

Narinder (centre) with friends.

Deadheading

There are times, I think, when Father knew me better than I knew myself.

Not long after I finished school and before I went on to college, I was anxious about not having anything to do: there was too *little* time to take on anything substantial – whatever that might be – and too *much* time to fill it with something frivolous. Father must have sensed my anxiety, because one morning over breakfast he suggested that I spend the next few days helping him out in one of his multiple 'side' businesses, a small trucking and hauling concern that ran mostly industrial goods from our hometown to various locations as far as 200 miles away.

'I just want you riding up front in the cab with the drivers. See how the business works. Maybe even tell me how to make it work better.'

'Do you want me to help loading and unloading?' As I recall, there were about a half-dozen vehicles in Father's fleet, with the cargo usually being trucked from a factory to a manufacturer or from one small business to another.

'As you like,' Father said, 'but mostly, just watch and learn. And,' he added somewhat cryptically, 'I'm certain you'll discover something valuable.'

Doing as he asked, for the first three days of sitting in a truck cab for eight to ten hours, I didn't discern anything valuable other than the fact that, to a man, each driver had an interesting life story to tell – from the youngest, who was not much older than I was, to the oldest, who looked considerably older than Father. It was not until we fuelled up before our departure from the last delivery on my fourth day on the job that it occurred to me that today – and on each of the prior three days – we were making fully half of the day's journey with an empty truck, save for the driver and myself. And then it hit me in a blinding flash of the obvious: *We weren't carrying a load home.* Yet our costs were fixed, whether the truck was empty or not. Which is to say that, for at least half the day, not only were we not making money, we were losing it. I've subsequently discovered that there is even a term to define this waste of resources: deadheading.

That evening at dinner, I decided to *not* share my insight with Father. Not until I had more to offer. After all, if I had figured this out, surely he had as well. In the meantime, he hadn't asked how my day went on any of the four days I was on the job. It was as if he'd completely forgotten the charge he'd given me. But *I* certainly hadn't. And in the days that followed my initial insight, I asked all our delivery customers whether they had anything to ship back to our neck of the woods, or if they knew of any other concerns that might.

Somewhat to my surprise, this early venture of mine into business yielded near-immediate fruit. And so it was that, in just a few days' time, as the driver and I were about to complete our first round trip with a nearly full truck, I announced to Father that our little business was possibly on the verge of doubling its income.

He smiled. It was the sweetest smile, my father's – a smile about the possibility of doubling his business, to be sure, but more likely, I'm convinced, a smile suggesting that whatever he had intuited about me had been confirmed.

Partition

It is the middle of the night, 15 August 1947, when I'm awakened by an insistent knocking on the door to my room in our family house in Dehradun. 'Kakaji! Kakaji!' our servant Gulu Mian calls out for me. 'The city is on fire! They're coming to kill us!' Gulu, a Muslim, had worked for my family for much of his life. He, his son, and his son's wife, all Muslims, lived in separate but attached servants' quarters.

It is the eve of Partition. The British are making their exit – the Empire can no longer afford this distant, occasionally recalcitrant colony of rascals, as Winston Churchill would sometimes refer to India's homegrown leaders. In the capitals of London and Delhi, lines have been drawn, new countries have been defined, and decisions have been made with little apparent regard for the human price to be paid. The Bengal Province of British India has been divided into East Pakistan and West Bengal, while the Punjab province where we lived has been sectioned off into West Punjab and East Punjab, with different religions and cultures assigned different territories. In the ensuing and frantic migration to their newly dictated homeland – and, in many instances, their escape from the murderous purges in the old – over fourteen million Hindus, Muslims, and Sikhs were displaced, the largest migration in human history, leaving 1 per cent of the world's population homeless, and as many as two million dead.

Gulu has every reason to be fearful. I pull on my shirt and

trousers. It is a warm night. When I open the door to reveal Gulu in his nightshirt, his face a dark smudge in the unlit hall, the heat hits my face like a wall. Behind me is the safety of my room and the last vestiges of my childhood; in front of me is the life of a young man in an India violently flipped upside down.

My father, mother, older brother, and sister are miles away in Kasauli, visiting my younger sister Davinder at the sanatorium where she is being treated for tuberculosis, a disease deadly at the time, and one that she had contracted caring for an older tubercular woman. All of nineteen years old on Partition Day, I am the man of the house, responsible for the safety of my younger brother Gurdev and sister Surinder, who are sleeping undisturbed through the night, as far as I know. But I am also responsible for Gulu's safety and that of his son and daughter-in-law.

'Come. Listen,' Gulu insists as I cinch my trouser belt tight. I follow him to the balcony on the second floor. It is there that I first hear the screams – animal screams like coyotes, emanating from the direction of the city about a mile and a half away. 'Muslim!' Gulu shouts, as if a Muslim's scream can be distinguished from a Hindu's or a Sikh's. But he is probably right. While there are a significant number of Muslims in Dehradun, they are far outnumbered by the Hindus and Sikhs, some of whom, in the days leading up to the official Partition, have taken it upon themselves to hasten the Muslim migration, turning it into genocide. But it isn't only Muslims on the receiving end of the violence. Every group, it seemed, has been caught in the madness and has become its victim.

Meanwhile, from the second-floor balcony of our house I can see pockets of the city in flames, sparks rising in the evening like fireflies, only dangerous, incendiary. I stand there next to Gulu, surveying the scene, our faces lit orange. And then from somewhere inside me (duty? responsibility?), the reassurance emerges, and I speak: 'I'll take care of you, Gulu. Your family too. No need to worry.'

'Thank you, Kakaji,' he says, 'I will tell my son and his wife.' And with that he is gone. But not from my memory, I realize, recalling a scene perhaps ten years earlier. I am still a boy, spoiled and foolish. I am playing with friends and suddenly feel thirsty. I am sitting only a few feet from the water tap with clean, empty glasses on a tray nearby, there for the taking. Yet, instead of pouring my own glass of water I call to Gulu working at some task across the room. 'Get me some water, Gulu!' I demand. Gulu looks up from his work. 'As soon as I'm finished here, I will bring the water to you,' he promises.

'Not soon enough!' I say, winking to my friends as if I had just proved to them my mastery over our servant. 'Not soon enough,' I insist. What I don't realize is that Father is in the next room and has heard everything.

'That's enough impudence out of you, Bino,' he says. With Father's appearance in the room, my friends scatter, leaving just the two of us standing by the water tap until Gulu joins us. 'Now, Bino,' Father orders, 'you are going to apologize to Gulu for your rudeness.' I remember murmuring an apology, one not sufficient for my father, who holds me by my shoulders and plants me directly in front of Gulu. 'Say you're sorry as if you mean it,' Father demands. And this time I do, with all my heart, asking forgiveness as well.

Watching Gulu pad away toward his room in his house slippers on Partition Eve, I wonder if he too remembers that earlier scene, one that has coloured our relationship forever, at least in my mind.

That night I sleep fitfully, never changing out of my shirt and trousers into my pajamas. The next morning, I am up with the sun. After washing the sleep from my eyes, I look in on my younger brother and sister, who are still asleep; all appears well. I see no movement from Gulu's quarters. Certain that everyone is secure, I ride my bicycle into the city.

Among the houses untouched by the night's mayhem are the smouldering skeletons of others. As I pedal through the city, the scent of burnt wood, tar and sulfur is pervasive. Every now and again, I see a bloodied corpse I assume to be a Muslim left in the street to moulder. Worse still, and by far the most horrific sight of all, is the front of a small school building where fifteen – I count them – young Muslim girls lay dead, one next to the other near the building's entrance, as if they had been executed by a firing squad. Not knowing what to make of the gruesome tableau before me, I stand transfixed: How could this horror happen in a place so near my home? And possibly, even committed by people I knew? Who perpetuated this heinous crime, Hindu or Sikh? Professional assassins, or scared or vengeful citizens?

Presently, two men arrive in a large pickup truck. They are both wearing turbans. Scarves cover their faces either to fend off the stench of death or to remain anonymous. Most likely, both. Methodically, as if they've done it before, they pick up each girl – one grabbing the legs, the other the shoulders – toss her into the truck bed, and drive away… to where, I have no idea. I doubt a dignified burial or cremation is likely.

The sights of the morning burning into my consciousness, I bike back home, feeling lost and vulnerable. How I wish my parents were here with me – Father in particular. Man of the house in his absence, I worry and wonder if I'll be called on to act.

I don't have to wait long to find out. By mid-afternoon, a crowd of about one hundred Hindus and a scattering of Sikhs have gathered in front of the gate leading to our house, only about 50 feet from the veranda surrounding the building and the entrance proper. A few of the men in the mob carry sticks, others torches.

'We know you have three Muslim servants in there, Sardar Ji,' one of the men yells, the most vociferous. 'Send them out and you and your house will be spared.' In the front room, Gulu looks to me beseechingly. 'Don't worry,' I assure him, the second time I have done so in just hours, now worrying myself precisely how I will defend him… and myself… and the others, for that matter.

Right then I spot the double-barrelled shotgun that Father keeps loaded next to the front door – to hunt birds and deer, he claimed, though I have never seen him do either. I pick the gun up – the first time I ever have done so – and with no idea how or where I summoned the courage, I wrench open the front door and stand on the threshold facing the growing, milling crowd. Suddenly, inexplicably certain of myself, I raise the shotgun, fix my aim at the gate, and call out to the threatening vigilantes: 'If any of you even *touch* that gate I will blow you to pieces!'

Responding to the threat, the murmuring crowd grows louder with shouts of 'Send out the Muslims!'

'Never,' I insist, now taking direct aim at the man with his hand on the gate latch. 'Take your hand from the gate,' I command. 'This is my last warning.' I'd never fired a shotgun before, never fired any weapon. Luckily, though, I'd been to the movies, seen the American Westerns.

Fortunately for all concerned, the few Sikhs in the crowd break ranks and try to placate the others. 'It's only Kapany's son,' I hear one say. 'Leave him be.' And then, 'Let's go.' Miraculously, or so it seems, what now looks more like a ragtag band than a murderous mob disbands, trailing away in small groups, possibly to wreak havoc elsewhere. But not here. Not at Kapany's place. Not today. I walk back into the house and set the shotgun down.

Of course, the woes and crimes brought on by Partition Day were by no means over, and even as green as I was, I knew so. The very next morning, in fact, a band of about a dozen young boys, all about ten years old, pried open the lock to a small toy manufacturing company – no more, really, than a large house with a few windows – about 100 feet from the end of our property. The toys made there were neither expensive nor elaborate, but were rather primitive, carved from wood.

Once in the factory, each boy grabbed what he could, two or three toys at the most. Just as the child-vandals were about to make their way higgledy-piggledy out of the place, a British military half-ton truck appeared on the horizon and braked to a stop right in front of the small factory. I watched, rapt, as six British soldiers jumped from the rear of the truck, quickly taking stock of the situation: the fleeing boys, the open door to the factory.

Without a word to his mates-in-arms, one of the soldiers took a knee, sighted his rifle, and shot the boy closest to him in the back. The toys he was carrying tumbled to the ground with him. Following a brief confab among the soldiers, the six of them climbed back into the truck and disappeared. About an hour later, a single, non-uniformed man driving another British military vehicle picked up the body and tossed it into the truck bed. He then picked up the boy's contraband toys and carried them with him to the cab, tossing them in through the open door and brushing off his hands: all in a day's work. It was the sheer normalcy of the gesture that was the most disquieting.

It wasn't that the British were *all* villains. Or villains *all the time*. Even at nineteen, I already knew that, over the years, the British had accomplished a great deal in India. Also, *for* India, which, fragmented into many small states as it was before colonization, was unable, or perhaps unmotivated, to achieve what the British did in subsequent years, and on a large 'national' scale: construct a network of roads and railroads and create an elaborate canal system, for example, not to mention provide a unified government capable of organizing and administering civil services. Even if most Indian precolonial state leaders had the vision to modernize their states, they had neither the resources to pay for their schemes nor the far-flung networks of trading partners for their goods.

No, my primary complaint with the British, during my days in 1940s India, was exemplified by their drawing of national borders that defined a new India but that were seemingly arbitrary, demonstrating little sensitivity to or understanding of the cultures that were affected.

That day in 1947, this insensitivity led to the murder of a single boy by a single British soldier right before my eyes, no doubt about it. A century earlier, in North India, it had led to the killing of thousands of largely unarmed, freedom-seeking Indians, when the British sowed the seeds of discord following the power vacuum left by the death of the great Maharaja Ranjit Singh in 1839, the last Sikh maharaja of Punjab.

Here was a man, a truly beneficent ruler, who controlled an Indian empire larger than the state of California. Highly enlightened and worldly, he brought in leaders, experts, and thinkers from across Europe and Asia to run the military and oversee various departments of government. So powerful and effective was Ranjit Singh's rule that the British knew that they could do nothing to change it. Not long after he died, however, the British were able to foment the disharmony and dissatisfaction that, in short order, set the stage for a series of bloody conflicts.

A few days after the shooting outside the toy factory, some of my Sikh friends and I got together. We were all about the same age, five of us in total. I'm not sure who brought up the topic – perhaps I did, still on a high from my experience with a shotgun – but one of us said, 'Partition is happening. Muslims are killing Hindus, Hindus are killing Muslims.' Sikhs, for some reason, were left out of the litany. 'There's crazy stuff going on. We should have a gun!'

'But you *have* a gun, Narinder,' Jasbir, my best friend, insisted. 'Yes,' I said, 'but it's a *shotgun*. It's too big, too cumbersome.' Even as I said it, I wondered, too big for what? And if we secured a gun, where would it reside? And who among us would carry it? And precisely *why?*

The answers to these questions were brought into sharp focus only a few days later when one of our group convened a special meeting at my house. The five of us were sitting on the veranda: three together on a couch, one in a sedan chair, and Jasbir on the concrete floor. 'I have a gun,' the friend who had arranged the get together announced. When we asked him how he came by it, he insisted he had simply found it.

'And bullets,' he added. The five of us were unnaturally quiet as he produced the gun from his jacket pocket. It was a small black revolver with a snub nose. There was nothing graceful about it. We passed the gun around among us. It was heavier than I'd anticipated. It was also loaded, a bullet in every chamber. No one, I noticed, dared to touch the trigger, or even the trigger housing. There were more bullets in a separate little box that we also passed around. For some reason, both the loaded pistol and the extra bullets found their way into my hands at the same time.

'What are you going to do with it, Narinder?' Jasbir asked.

And again, for some reason I could not discern, this decision – like the pistol and the bullets seconds before – fell into my hands. Unlike the shotgun, which looked and felt antique, this gun felt new and dangerous. 'Unload it!' I said it without thinking, seeking only to diminish the danger. 'The first thing I'm going to do is unload it,' I said. Heeding my own instructions, I snapped open the cylinder and shook the bullets from it into my free hand. I then distributed them, along with those in the box, equally to the others. The tension on the veranda seemed to abate somewhat.

'I'll keep the gun,' I announced by way of a plan, 'and you'll each keep some of the bullets. And if I need them, or you need the gun, we'll each know what to do, where to go.' Everyone seemed satisfied. No one, when faced with the burden of having a loaded pistol in his possession, wanted the responsibility, it seemed – not even any one of us, a bunch of cocky nineteen-year-olds. Whatever tension was left on the veranda dissipated as I held the gun in my hand, jokingly taking a bead on Jasbir.

'Hey, Narinder, don't mess around,' he said, raising his hand before his face as if to fend off an unaccounted-for bullet.

What neither Jasbir nor I – nor any of the others – knew was that there actually *was* one live cartridge left in the chamber. And had I pulled the trigger while the gun was still pointed at Jasbir's head, I would have killed him, without a doubt. And my life, full of hope and promise back then, would have borne no resemblance to the seventy rich and fulfilling years I have lived since that time.

Unaware of any of this, I turned the barrel of the gun down to the concrete and squeezed the trigger. The retort was incredibly loud and horrifying, as small shards of concrete flew every which way and as a bullet ricocheted and disappeared. The five of us stared at each other, not a word among us. (I'm not sure what ever later happened to that gun.)

A sad afterword: Jasbir died relatively young, in his sixties. He had married, raised children, became a successful businessman in India. Satinder and I visited them whenever we were in the area and we remained friends over the years, the four of us. One day, three young men answered Jasbir's advertisement for a car he was offering for sale. The four of them took a test drive into the hills where the boys killed him and stole the car. They were never apprehended.

In the days, weeks, and months that followed the Partition, I was curious about what the ultimate outcome would be. How would the newly formed countries change? How would the cultures themselves change? Of course, years would go by before there was any definitive answer. And in Pakistan, in particular, the outcome seems still to evolve daily. But, as the summer of '47 wore on and I had time on my hands, I often biked – alone or with my friends – to some of the not-too-distant refugee camps in Punjab where immigrants were living out in the open under the summer sun, mostly Hindus and Sikhs arriving from Pakistan, poor, impoverished, and road-weary. I saw other camps, too, many with Muslims leaving Punjab, heading for Pakistan. At the time, and since, it all made precious little sense to me: millions migrating, more than a million dead, and for what?

Two years later, in a literary journal I founded and edited at my college, where I was a member of the Sikh Students' Literary Society, I wrote that young men should be 'protected' from the follies and vices of manhood – politics and war, in particular – as well as from the temptations and dangers of the times. And that instead, they should be schooled in deep learning, true culture, fine art, and pure religion, so as to 'enrich them with the right sense of understanding and power of free thinking…'

In that same issue, my short story, 'Innocent,' railed against atrocities committed in the name of religion: 'the best armour in the world and the worst cloak.' In it, I described the worst and final day in the life of a farmer and the bloody massacre of his family.

These two entries, more than anything I can say today about that period of my life, provide the best insights into my frame of mind some seventy years ago, in the period surrounding Partition and the time when I ceased being a boy.

Seated, from left: Jasbir Singh and Narinder with friends, picnicking on the bank of the River Yamuna where it emerges from the Himalayas at Paonta Sahib, Himachal Pradesh.

PART II

LEARNING

Facing page: Watercolour on paper, artist Sumeet D. Aurora.

Obsessed

'Light can only travel in a straight line.'

The professor's words roused me from a gentle slumber. I shook my head to clear it, and looked around. I was one of about thirty students in an introductory physics class at a regional outpost of my college, a short bicycle ride from my home, where I lived with my parents, brothers, and sisters. The professor was young, self- assured, and good humoured, one of my favourites. He stood at a podium on a raised platform before us, a wall of blackboards with indecipherable scribbles behind him. It was 1947.

'What, Kapany? Have I interrupted your nap?'

'Just thinking,' I said. But I was more than just *thinking*. I'd already *thought*! And I was certain that he was wrong: that light couldn't only travel in a straight line.

About napping, however, my professor was correct. In fact, I'd been napping for nearly all of my first three years of college, drifting like a somnambulist from course to course, waiting for the one that would awaken me. So far, however, the only one that moved me at all did so literally: classical dance, in which we young men danced in the glare of a bright light that threw large shadows of us onto a massive white screen at the rear of the stage, drawing even greater attention to my gracelessness.

Truth be told, I was not cut out to be a professional classical dancer. A relatively tall, slender fellow with a scraggly beard, who wore either a black or a blue turban, I favoured tennis and field hockey, and attended soccer matches with friends. I was, in other words, an altogether normal Sikh student in my early twenties. There were some (I was told) who even found me good-looking.

I was running out of time, however, to find my place in the world. Graduation was just months away. Fortunately, I had a Plan B: to follow in the steps of my father, who had served in the Royal Indian Air Force. With the end of school looming, I applied and was accepted, though my mother insisted that I was far too entrepreneurial for the armed services and would quickly become impatient with military life. Still, it was my ace in the hole.

It was also an eminently practical thing to do, though back then I wasn't at all certain where my true nature lay along the scale between the practical and the theoretical. Was I a practical theoretician? Or a theoretical practitioner? Or perhaps I had no fixed place on that continuum at all, but was instead a chameleon, able to change colour as the need arose. Or maybe my colour remained constant, and I just slid, barely noticed, along the scale.

'Light can only travel in a straight line.'

So spoke my favourite professor in that introductory physics classroom in 1947.

'No! No, it doesn't,' I wanted to shout, believing it with every fibre of my being but being far too well brought up to challenge my professor in front of others. Or to reveal myself the fool. Besides, it was just this morning that my mother – who over the years had expressed only passing interest in my near-four-year sleepwalk – had said, 'Your father and I have the highest hopes for you, Bino.' But that was *then.*

'Light can only travel in a straight line.'

...and *now*, feeling inexplicably energized, I sensed a light go on in my brain. No ordinary light, it seared me to the core, reflecting and diffracting from lobe to lobe, and it definitely was *not* traveling in a straight line. And just like that, sitting in an undistinguished classroom in North India, on the cusp between waking and sleep, I realized my life's work: not only to prove that light could travel every which way, but also to put this simple physical truth to work in amazing ways. And with that illumination, I awoke – and became obsessed.

Narinder at Dehradun.

More Mentors

Following graduation, my obsession to discover as much as I could of the paths and power of light took me to an ordnance factory in Raipur, where I was fortunate enough to work for my second mentor, another fellow like Mr. Zachariah, who placed considerable faith in me and helped me explore the optical universe.

His name was Mr. Shahni, and he was the chief executive of the ordnance factory that provided me with my first full-time employment. For a curious young fellow like myself, the factory was the perfect place to apprentice in the design and manufacture of all manner of optical devices, including microscopes, binoculars, range finders, and telescopes.

I remember our first meeting. We were standing across from each other in a quiet corner of the factory floor. Mr. Shahni fixed his eyes on mine for a number of seconds before he spoke, as if trying to *read* me – with me feeling, back then, that there was precious little yet to read. Also, I felt that my job was quite tenuous. I'd been officially hired a few days back and was drawing a salary, but no one had yet told me what I'd actually be doing. That didn't seem to faze Mr. Shahni. 'What would you *like* to do here?' he asked, 'If you had your druthers, that is.'

I may not have had a clear understanding of who I was, but ever since my professor's assertion that light couldn't be bent, I'd known exactly what I wanted to do. So I told him. 'I want to learn everything I can about light. About optical instruments.'

Mr. Shahni smiled slyly. 'Everything? Everything, large and small?'

Was he toying with me? I didn't know, but I repeated my request. 'Everything.'

'And what job title would you like?'

'I don't care,' I said. And I didn't. Not really. I didn't need a position or job title; all I wanted was to learn, and I also told him so.

'I like you, Kapany.' He stood, extending his hand for me to shake, and I took it. 'And I'll do better than that.'

Better than what? I wondered. Our first meeting seemed to be moving along at a good clip, but I wasn't sure I was tracking it. 'I don't quite understand,' I admitted.

'You will.'

A few days later a letter arrived from Mr. Shahni. In it, he accorded me the title 'Supervisor' and specified that I would take part in a two-year training program that required I work hands-on in every department of the company – glass cutting, grinding, polishing, and edging – as well as learn all the mechanics involved in the manufacture of a wide variety of optical devices.

By the end of my specially designed apprenticeship, I could literally build a binocular from scratch. Whether he was aware of it or not, Mr. Shahni had set me up for life. Under his tutelage I became a student of light and its properties, a skilled technician of optical devices, and, in the years to come, a mentor to those I would teach, work with, and eventually employ.

Still, by the end of the second year of my tutelage, I was growing tired of the repetitiveness of the work at the factory and was eager to move on. My study there of basic optics had whetted my appetite for more challenges.

I found those challenges in 1952 at Imperial College in London, which provided the research opportunities I was seeking and, as I was soon to learn, was a veritable playground for those equally obsessed.

As luck or fate (or both) would have it, it was here that I met and studied with Professor Harold Hopkins, who, in short order, became my third mentor, and whose support would forever change my life. The insights and work that emerged from our relationship changed *his* life as well, I believe, in a form of transference I've since found typical between mentors and mentees. The idea that obsessed us both was that light and images could be transmitted through flexible axes: Light could, in other words, be bent.

Satinder

And yet, as drawn as I was to explore this idea, by the end of my first semester in London, I had found a second love that was equally compelling. It all began when Jimmy, my Parsi friend from India, asked me to go on tour with him and his student dance troupe and give demonstrations of traditional Indian dance over the Christmas holidays. I turned him down with nary a moment's thought. After all, aside from my brief flirtation with classical dance and shadow dancing as an undergraduate, I had minimal dance skills. Nor was I particularly partial to public performance of any sort. And *demonstrating* something I couldn't do myself… well, that was completely out of the question. 'I can't do it,' I insisted. 'Besides, I need all my free time to study.'

'You won't have to actually *dance*,' he countered. 'I just need you to strike a gong at key moments in the performance.'

'A gong?' Was this truly going to be the sum total of my participation?

'There's more for you there if you want it, Narinder. But otherwise, yes! When I give the signal, you strike the gong.'

I was about to turn him down once and for all when a few female members of the troupe appeared at the doorway to Jimmy's office, where we were sitting. They were in Indian dress, their bracelets jangling, their laughter rich and inviting. Standing slightly apart from the others, her look both demure and assertive, humble but regal, was a stunning young woman. A necklace of opals, the light twinkling in the polished stones, was wound twice loosely around her neck.

Jimmy introduced the women one at a time and I shook the hand of each, forgetting their names in an instant, except that of the woman with the softest hand and, by far, the firmest grip. She was the woman with the opal necklace, Satinder, who met my gaze unflinchingly.

'I've heard of you,' she said to me. 'You're the glass man.' She was also a student at the time, studying English literature. 'And you'll be traveling with us?' she asked. 'On the tour?'

What little resolve I had to remain in London over the holidays disappeared like a fast-receding tide. 'Yes,' I told her. 'Absolutely! I'll be on the gong.' I mimed striking a gong and made the appropriate sound effect.

'Wonderful,' she said. 'And you'll be dancing, too, of course?'

I scoured my memory for past traditional Indian dance performances, observed or partaken in, but couldn't come up with any. 'Of course,' I heard myself saying, speaking to this woman who would be my wife in less than three years, and my life companion for sixty-two.

Apparently thinking the deal had yet to be struck, Jimmy continued to try to convince me to go. 'Come on, Narinder, it'll be fun. Take you away from your studies. Think of it as a vacation. I'll even cover all your expenses.' To this day, I have no idea why Jimmy had me in mind for his gong man; I'm only eternally thankful that he did.

Besides, it turned out to be a great road trip – the best. There were ten of us in all, five men and five women, plus Jimmy, with plenty of good cheer and alcohol to be had as we travelled from venue to venue by train and boarded in cheap hotels, with Jimmy always settling up for us.

Not surprisingly, the performances weren't at all well-attended, often with more empty seats than full ones. But, as Jimmy assured me, it was fun – and I met my lifelong love. I even learned a few traditional Indian dance steps.

What ultimately sealed the deal for Satinder and me as a couple, however, was not my dancing prowess, but my pharmacological powers, my ability to cure the horrible headaches she was sometimes afflicted by, and that she confessed to having on the first day of the tour.

'I have some special tablets to give you,' I told her. 'I, myself, have already experienced their amazing curative powers. And if you'll permit me, I'd like you to avail yourself of their magic.'

Dubious but ultimately willing, she accepted what were a half-dozen tiny Sen-Sen – a popular remedy back then for bad breath. And on my instruction, she held them under her tongue until they dissolved. It was a process that took about three minutes, and when it was through, she looked up to me as if I'd performed a miracle.

'Where, oh where, will I ever get those tablets again?' she implored.

'I have quite a supply,' I assured her. And it wasn't until 2013 when Sen-Sen was taken off the market that I ever had to turn her down.

Narinder and Satinder Kapany.

My Wee Scottish Sojourn

And yet, despite my growing love for Satinder, and after only a year of graduate study at Imperial College, I had a hankering for the 'real world' again. So I took a position as a researcher at Barr & Stroud, a renowned optical engineering firm in Glasgow that designed and built optical devices for the military.

I met in London with the director of the company, Mr. Morrison, and we hit it off almost instantly. As I recall, he even recommended a rooming house where a number of other Barr & Strouders boarded. The proprietress, a Mrs. Reed, was a short, plump lady who took a liking to me, for some reason. And while she served all the other boarders haggis and other Scottish specialties less-than-appealing to me, she made me delicious potpies and other treats. Better yet, when Satinder came to visit, Mrs. Reed treated us like a honeymoon couple, seeing to our every need – though, in truth, we had few except each other.

One night, without checking first with Mrs. Reed, we decided to have dinner in a pub, unaware that in 1950s Scotland, women were not welcome in pubs in Glasgow. Not to mention many of the pubs in the downtown were also host to gangs and the criminal element, not safe places for a nice young Indian couple, in any event.

I should have guessed that something was seriously wrong from the moment we walked in. Seated halfway into the room and directly across from the entrance was a group of hooligans that I recognized from newspapers as Razor Slashers, so called for the razors they sported sewn into the bills of their caps, strictly for the purpose of committing mayhem. The instant I registered what I saw, I felt fear grip me: not the fear of standing up to the mob on Partition Eve. That was a cool, rational, sensible fear that, for some reason, I knew I would survive. But here in this smoke-filled bar reeking of stale ale and unwashed male bodies, fear was like a knife in my belly. I turned to Satinder, who tried bravely to meet my gaze then broke out in tears.

Uncertain of what to do, for some reason we persisted forward, found a table in the back, ordered some food from items scratched on a blackboard above the bar, and seriously caught the attention of one of the Razor Slashers, who ambled over to us, hand on his lethal cap. Arriving at our table, he mumbled something menacing in a Scottish dialect I didn't understand. His cohorts chuckled, a mean burst of crude laughter. I heard a nearby chair scrape the floor.

'We have to leave here now,' I said to Satinder, standing. 'Yes.' Her voice broke as she stood and grabbed my hand, so hard I thought she would crack my knuckles.

'But not conspicuously,' I added, not even knowing what I meant by it.

'Yes,' she agreed again, as we both turned and, feigning as best we could a look of nonchalance, backed out of the pub and into the rainy Scottish evening. It was a downpour, in fact, pure and safe. Glasgow rain never felt so good.

Not long before I left Barr & Stroud, I was asked by a Scottish mathematician who worked at the company to be the guest of honour at an upcoming birthday celebration for Robert Burns, a national hero and literary idol. Which, reasonably enough, led me to the question, 'Why me?' I mean, here I was, an Indian, a Sikh, a man of science rather than a man of letters, and, most significantly, only in my mid-twenties. 'Why me?'

The mathematician laughed but offered no explanation. Not to be discourteous, I accepted the invitation, along with the request that I prepare a few remarks.

It was only after that cryptic invitation, and after scouring the library at local bookstores for Burns's works that I realized we didn't speak the same language. I couldn't read a word of his poetry. Not a word. Surely the mathematician must've known this. At some point, I even suspected that I was a victim of a cruel prank. But no, the mathematician seemed earnest enough, so I determined to give the speech my best shot.

First, culling some biographies, I wrote some opening remarks that celebrated Burns's life and literary accomplishments. Next, without knowing what the words actually meant, I picked a few of his more highly regarded poems to present, but not before having had them read aloud to me by one of my Scottish boarding house cronies, and memorizing the correct pronunciation. Finally, I had all the poems that I would read printed out on separate sheets that I left at the places of all the banquet attendees – about 200 Scotsmen – so they could follow along with my reading.

As if all that weren't enough, when it came time for my presentation, I stood proudly before my audience wearing my altogether expected turban as well as a totally unexpected kilt. It was the perfect sight gag and had the audience clapping long after the obligatory welcome. And if I dare draw conclusions from the audience laughter in all the right places, as well as the standing ovation that immediately followed my remarks – revealing many more kilts – my talk was a roaring success.

Not long after – I worked at Barr & Stroud for only a few rainy months – I received a letter from Professor Hopkins telling me that he had taken my ideas for exploring light to the Royal Society and that they were very excited about my work. So much so that they were offering me a full scholarship, including a small stipend for personal expenses, as well as entry into the Imperial College's PhD program in physics. 'Come back, Kapany! Come back, Narinder,' Hopkins wrote, and underscored 'Come back!'

It was an offer simply too good for me to refuse, the very best in my young life – so I didn't. Of course, I checked out with Mr. Morrison, who seemed very happy for me and wished me Godspeed. About a decade later, in the early 1960s, when I already owned a company of my own and was looking to acquire companies abroad, I paid a visit to Barr & Stroud, where I found Mr. Morrison still on the job. 'Why did you go to America?' he asked at one point. 'Why didn't you just come back here? Was is it about money?'

I assured him not, then thought carefully about his question. 'You know, I just wanted things to change a lot faster than they did here.' As soon as I said it, I feared that I'd insulted him. I even asked if I had. 'No, no, no!' he demurred. But then he asked what I'd meant.

So I said, 'Look, I was here ten years ago and haven't been back since. Still, I'll bet I can sit here and tell you exactly how every workspace in the factory is configured – precisely what pieces of equipment are where. And then we can walk over and see if everything is still in its place.' Which we did, the two of us together, proving my hypothesis, as it turned out.

'I see,' he said, at the end of the tour without my having to prompt him, 'I see.' He flashed me what appeared to be a bittersweet but knowing smile.

Narinder at Hyde Park, London, 1954.

Rainy Day Woman

Upon my return to London in 1952, I'll admit, I was lukewarm at first about the PhD program. Not merely about the Imperial program, but about *any* PhD program. And for a good, if misguided, reason. In 1950s India, those earning PhDs had the reputation of being impractical do-nothings. That was not for me, I originally thought. But I was quickly disabused of this notion by my colleagues in the program.

And thank goodness for the small stipend. Back then it was virtually impossible, if not illegal, to exchange rupees for pounds, both in Britain and in India, so there was no way my parents could help underwrite my studies. This made for some meagre meals in the oversized sarcophagus that served as my tiny one-bedroom, five-story walk-up flat in South Kensington.

Still, I had a hot plate, a couch that could be converted to a bed – and most often was – and a sink that I used both for washing my single dish and brushing my teeth: all a PhD candidate bachelor could hope for, and more. As for the stipend, it usually arrived on the first of each month, and I had to budget carefully so as not to run out of cash before the month's end.

Once, in fact, right after arriving back in London, I was on a date with Satinder. We were on our way to the cinema a few miles down the road, and all I had in my pocket was the exact change to buy two tickets. Suddenly, as so often happened in London, it began to rain.

And with the first drops, my wife-to-be urged that I flag down a taxi. Cash poor, I didn't have enough money for a taxi *and* the cinema. What was I to do? I suppose I could have confessed my lack of funds, but, doctoral candidate or not, I had my pride. So instead of hailing a cab, I said, 'Rain! Isn't it wonderful? Don't you just love walking in the rain?' A lesser date would have protested. But not Satinder. Instead, she miraculously produced a small umbrella from a large pocketbook, opened the umbrella, grabbed me by the arm, and the two of us made our way purposefully – and happily dry – to our destination, my finances and pride intact.

Scenes from the Kapanys' life in England.

Re-obsessed

'I'll need a lab,' I told Professor Hopkins on the first day of the new semester. 'One that I can darken completely, one that…' We had just left his office and were headed I didn't-know-where.

'Of course. *Of course*, you need a lab. In fact, I have a place all picked out for you,' he enthused. 'It's going to be very much to your liking, Kapany. To your very specifications, in fact.'

On the practical/theoretical continuum, Hopkins came out solidly at the theoretical extreme. Or was it the practical? Like me, Hopkins could vacillate between the two in a single sentence. Unlike me, he was a communist. And this was in the early 1950s, when it really meant something, particularly in places like America. But it never interfered with our work or our friendship.

We walked into a huge structure, then directly into a room with at least a 15-foot ceiling and large casement windows that ran almost the full length of the building on the right and the left. Dwarfed by the size of the room were at least thirty lab tables, many bearing lab apparatuses. A fellow of about my age was working at the far end of the room. He waved to us and then disappeared through a small back door.

Hopkins and I continued through the room and stopped some 20 feet short of the door. 'Here it is!' he said, pointing to the empty far-right windowless corner. 'Now go and do what you will with it, Kapany!'

'But,' I said stating the obvious, 'there are no walls, no door, no equipment here – too much light. There's no lab at *all*,' I finished.

'Only you know exactly what you want. So build it,' he said, handing me my own key to the building, the fob already bearing my name.

So I did, forwarding my bills for materials and equipment rentals directly to the Physics department and only asking for help when I absolutely had to, usually from one of my fellow researchers who seemed more than willing to take a break from his work – clearly less obsessed than I.

Using the two walls that formed the far corner of the building, I framed in two additional walls, each about 15 feet long, to create a room of approximately 225 square feet. I then roughed in a space for a door, and built what was essentially a lid that I nailed into the tops of the walls about 10 feet high. The final touch was a small two-sided sign for the door: one side that read 'Come In,' the other that read 'Do Not Open: Testing.'

Still, it would be some months before I actually ran any experiments. My plan at its most basic was to use individual, hair-thin glass fibres, identical in length and formed into a bunch that could be contoured slightly. I'd then shine a light through one end and determine whether it was emerging from the other. The problem was, I had no fibres, and there was only one company in all of England that was producing them in any quantity: Pilkington Glass in St. Helens, Lancashire. Pilkington was England's pre-eminent glass manufacturer – and a full three-hour train ride from London.

Professor Hopkins and I decided to make the trip together. Uncertain whether Pilkington was using only the highest quality optical glass – and not inferior green glass – in the manufacture of their fibre, I brought a good-sized chunk – about 8" × 10" × 3" – of polished, optical-quality glass along with us.

The train underway, Professor Hopkins was the first to offer up some conversation. Donning his more theoretical persona, his talk was of politics and the working man. Though I'd heard it before, I listened with interest. After a while he fell silent and I chimed in nervously to fill the gap in the conversation.

'Let me ask you this… You say *you* are *there*. And you say *I* am *here*.'

'Yes,' Hopkins ventured warily, though his look spelled amusement.

'Yes, well, how do you know that's true? How do you know that this is exactly what is happening? Why isn't it just a figment of your imagination?'

Hopkins chortled. 'Aha! So you're a philosopher as well as a physicist?'

'I try,' I began. 'I try not to be just one thing.'

'I see. Not only a man with a turban who shows up one day in a hired lorry filled with lumber and a serious-looking tool belt; you are a deep thinker, as well.'

'Not too deep,' I confessed.

But he ignored me. 'About your questions, Kapany. Have you ever had a toothache?'

'Yes, when I was a young boy. It was the worst toothache ever.'

'Good. Now when you had that toothache, was there ever any question about whether you were *there*, or not? Even the slightest doubt of it?'

Recalling the day vividly, the pain and my yowling, I offered, 'No, no question at all.'

'Well then, there you have it! Next question!'

At Pilkington we were given a guide to show us around the work floor. Of greatest interest to us, of course, was the company's fibre-making operation. Here, multiple molten fibres were extruded through pinprick-sized holes in the bottom of a red-hot furnace filled with molten glass. The fibres were then cooled and spun onto large spools from which they'd eventually be unravelled to be made into glass 'yarn' elsewhere in the plant.

The three of us stood briefly transfixed by the process, shadows playing over our faces, thrown by the flames firing the furnace. No one said anything for a few seconds before I unwrapped the optical glass I'd brought and held it out to our guide. 'Can you make us some? Some fibre, that is,' I asked, 'using this?' I offered the piece of glass to him.

The man eyed the optical glass dolefully, seemingly turning it down. Had I somehow offended him, suggesting that Pilkington's raw glass stock was in some way lesser? Defective?

'You needn't have brought your own,' the man pronounced, 'Pilkington's glass fibre will surely be quite adequate to the task, whatever your purpose.' I explained to him how I intended to use it, how every strand had to be crystal clear, and why only the highest quality glass would do. I held out the chunk of optical glass a second time and this time he took it. I'd have my spools of fibre, he assured me, as quickly as they could be scheduled in Pilkington's work flow.

Two months would pass before the spools of fibre were delivered to my lab, with much brouhaha. Virtually everyone working in the building had some sense of what I was up to, and they all knew I was expecting a large parcel from St. Helens.

Finally, with the arrival of the Pilkington fibre, I was ready to begin my work in earnest. It was a labour-intensive and wildly time-consuming task. In order to transmit a visually coherent and complete image through a fibre assembly, it's essential that each of the fibres that are part of that assembly be packed accurately, adjacent to each other, with each individual fibre carrying a single spot of the image. To achieve this perfect alignment, I laid out the pure Pilkington fibre in continuous lengths on a circular drum, itself an arduous and mind-numbing task that required 50,000 perfectly aligned rotations around the drum – the number of strands that I determined would constitute a single, complete fibre bundle that light could pass through while following the contour of the drum. Still, only months since arriving back at Imperial, I was ready to perform my first, and most basic experiment: to project a complete image through a curved, continuous alignment of fibres.

It was late one afternoon that I made my last measurement, the door to my lab still open, the room flooded with light from the nearby windows. I gave thought to ringing up Professor Hopkins to witness the test, but then thought better of it. By the time I pondered and re-pondered the notion of inviting him, teatime had come and gone and the dinner hour was upon us. I suppose I could have put the fibre to the test right then. All I needed to do was to darken the room, set up a light source at one end of the bundle, and then check whether the light came through to the other end and at what intensity.

But for some reason I still cannot explain, back then, more than sixty years ago, I wasn't quite ready for the result, whatever it would be. So, I shut off the lights in the lab, locked up, and went home to my sarcophagus where I made myself a hearty student's supper of canned beans and week-old bread. That night, my dreams were of bright, pure light without a single errant artefact.

And so it was that, filled with hope and expectation, I returned to the lab not long after sun-up. With little fanfare, I set up my light source at one end of the fibre bundle, turned off all the lights in the room, put up the 'testing' sign, closed the door, and in pitch darkness flipped the light source switch, and… nothing!

I switched the light off and then on again. Still nothing. I repeated the process a number of times with the same outcome each time. I thought back on that red-letter day in India and my professor's words that gave rise to my quest:

'Light can only travel in a straight line.'

I thought briefly about my family and my adventure with Jagdish stealing fruit from the maharaja's orchard. After sorting through a few more random thoughts, I fixated on the primary one: Professor Hopkins – and how glad I was I hadn't invited him to the lab for the maiden voyage of the *HMS Bino*. Still, as the morning wore on, I realized that I needed to pass the news on to the professor. Finally, around noon, I did. When I reached him, he seemed neither disappointed nor perplexed. 'Not to worry, Kapany, it's just a matter of time. And,' he added, 'you've got plenty of that.' He was then thirty-four to my twenty-five.

After much consideration that day and well into the evening, I determined that there was nothing inherently wrong with my hypothesis. Instead, and far more likely, there was something wrong with the fibre, some flaw in its manufacture, a flaw in the glass itself, though it was hard to believe the optical glass I'd nursed all the way from London to St. Helens was the culprit.

So, without giving it too much more thought, I took out a diamond-blade saw and cut a few inches off the bundle with the hope that the segment remaining would be pure and would transmit the light. Once again, I darkened the room and switched on the light source. Once again, nothing. So I tested the smaller piece I'd discarded, this time with the hope that it was only some fibres in the longer bunch that were somehow flawed or impure. Still nothing.

On the cusp of losing heart, I performed this cutting-down-to-size-and-retesting procedure a number of times, gradually coming to the conclusion that Pilkington never used any of the optical glass I'd brought from London. I recalled the sour look on our escort's face when he first turned down the optical glass. And the put-upon look when he finally accepted it. He'd probably tossed it directly into the rubbish bin. As for the spools and spools of glass that Pilkington had sent, they were probably made from old beer bottles. No wonder there was no light!

I was desperate. Still, I felt it was worth another try. This time, though, I sawed off enough fibre to leave only about two inches on the drum. With some trepidation, I set up the light source. Once more I affixed the 'testing' sign. With even more trepidation, I shut the door and turned off the overhead light, leaving only one more function left to perform. I hesitated for a mere second, then switched on the light source. Light streamed along the contour of the fibres and out the other end.

Later that day and on my request, Professor Hopkins appeared in the lab. In the time between my first successful test and Hopkins's arrival, I affixed a black mask with the cut-out word 'FIBRE' to a lens that I set up in front of the business end of the remaining Pilkington bunch and beamed it to a makeshift projection screen on the far wall.

'Well, Kapany, what do you have to show me?'

I pulled up the single chair in the lab, placed it with the best sight line to the screen, and urged him to sit.

'Ready?' I asked.

'Yes, yes,' he said, feigning impatience.

I closed the door, shut off all the lights, and in a single, final effort, I switched on the light source. And there it was. On the screen. Like an optometrist's chart but with only a single line of large letters, as clear as could be: F… I… B… R… E…

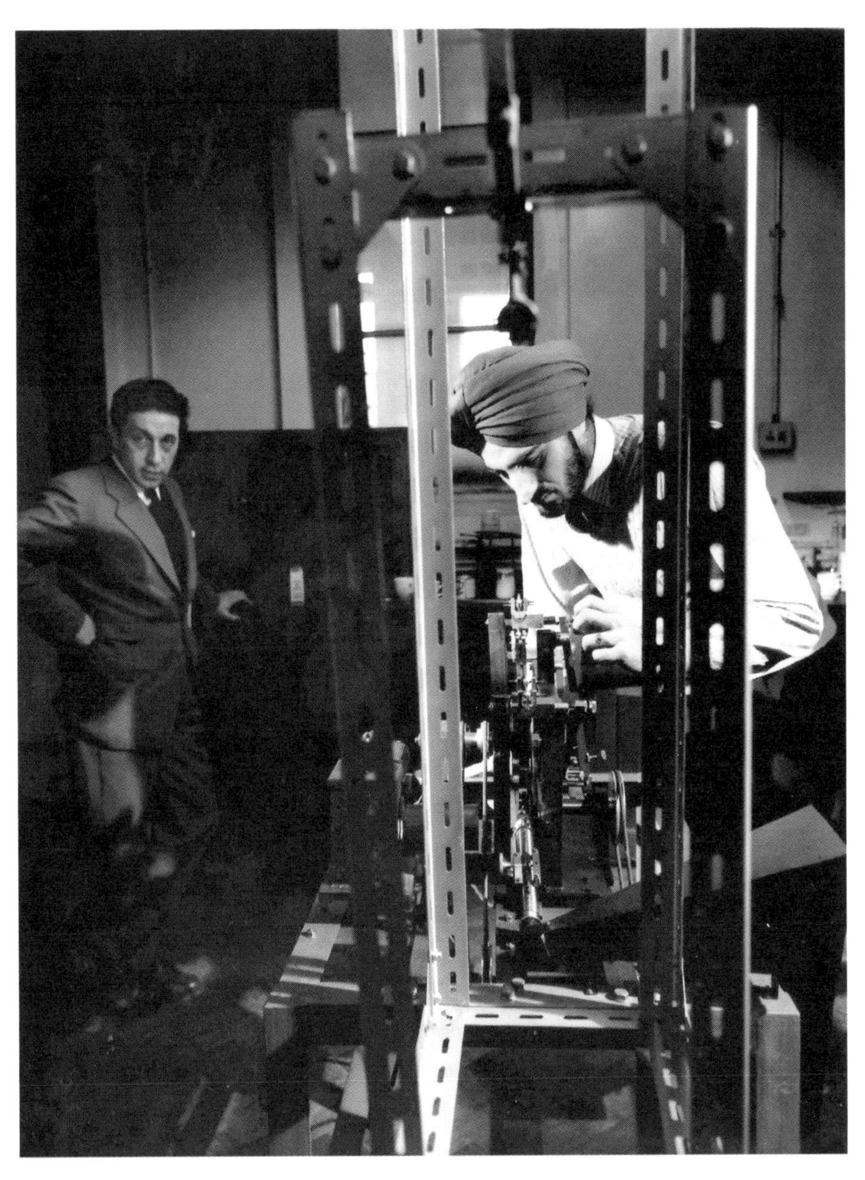

Narinder, then twenty-seven working in his research lab at Imperial College London, 1952.

PLEASE DO NOT ENTER

Popular Mechanics – Sept '55

Rope 'Scope

GLASS FIBERS finer than human hair make up the chief part of an optical instrument that can see around corners. The fibers are aligned in a rope bundle. By looking along the axis of the fibers, you can see an image at the other end, no matter how the rope is looped or twisted. Doctors may use it in internal examinations of the human body. Scientists could observe radioactive materials shielded behind lead walls and engineers could use it to investigate concealed parts of complex machinery. Known as the Fibrescope, it was developed by Dr. H. H. Hopkins and a 27-year-old Punjabi, Dr. Narinder Singh Kapany, at the Imperial College of Science in London. The simple instrument may replace expensive optical systems which are bulky and inflexible.

Even when a loop has been tied in the fiber, image at one end can be seen through a lens placed at the other end. Below, Dr. Narinder Singh Kapany adjusts the machine which aligns the glass fibers. Kapany, a 27-year-old Punjabi, is co-developer of the Fibrescope

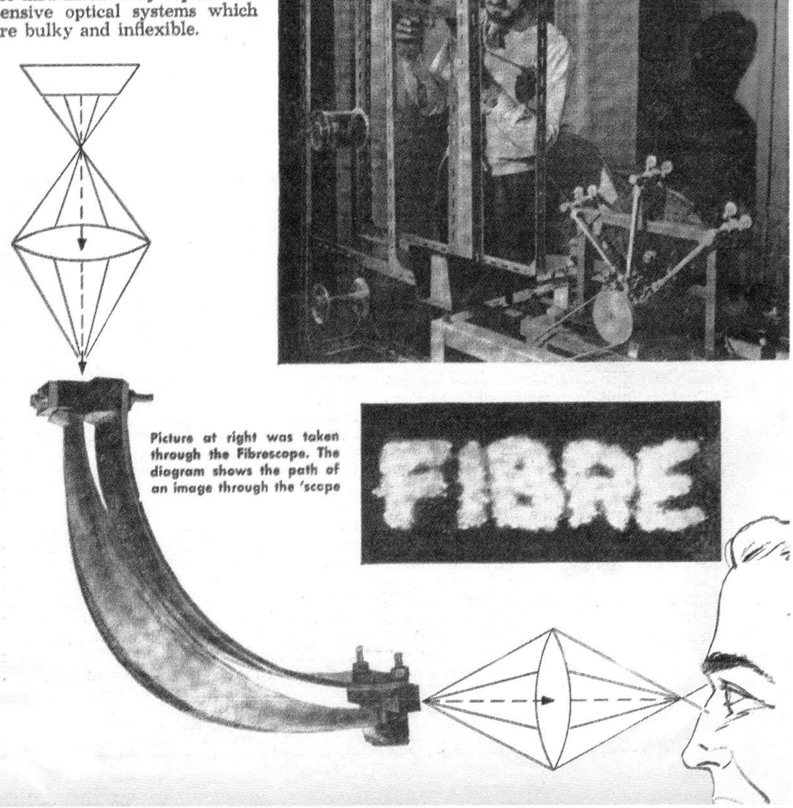

Picture at right was taken through the Fibrescope. The diagram shows the path of an image through the 'scope

Article about Narinder's 'Fibrescope' in *Popular Mechanics*, September 1955
Facing page: Narinder in his research lab at Imperial College.

London Snapshots

My actual time in London – from 1951 through 1955, excluding a few months for my Scottish sojourn – was relatively short when compared to the years I spent in India or California, but those London times figure large in my memory and on my emotional landscape. Perhaps it's because I did my formative scientific work there. Or because it was the place where I met and married Satinder. I still keep a flat there, in any event, and am a frequent visitor. There's just something dusty and old and, at the same time, exciting and new about the place. And I experience all four aspects whenever I visit.

Of my early times there, I remember, in particular, a Christmas celebration dinner at the grill just off Piccadilly Square. Satinder and I and a number of friends, toasting holidays past, holidays present, and any other holidays we could muster. The 'pop!' and froth of champagne bottles; I dare not venture how many or how loud we were. We abandoned the restaurant when we spied the first few flakes of snow cascading past the window to fall softly onto the square, the blurred glimmer of streetlamps projecting through their frosted lenses. Maybe eight of us in all, romping through the fresh snow, high, tipsy, and drunk with freedom and possibility.

Also memorable during my London stay was a February morning in the early 1950s. I'd arrived the year before, travelling from India to England by boat, having boarded in Bombay after staying with a cousin for one night. Knowing that I was heading off to a foreign city with but a single Sikh temple, she gave me two improbable, but not altogether impractical gifts: a tin of Indian pickles (Lord knows if I would otherwise have been able to find them in that backwater, London!), and a bright pink turban – both gifts I accepted graciously, though, as a point of information, the only turbans I ever wore were black or dark blue. Nesting the tin of pickles in the turban, I stuffed both into my baggage.

A year passed and Britain's heralded rains did not disappoint, leaving me one morning with a half-dozen black and blue turbans too damp and uncomfortable to wear, though I'd left them overnight on a shelf above the stove to dry out. It was then that I remembered the pink turban, still in my suitcase beneath my bed. I put the pickle tin on my tiny kitchen counter and the turban on my head. I looked ridiculous in the mirror, even more ridiculous the second time I glanced at myself – my misguided vanity speaking, to be sure. I was being too sensitive, I decided. Besides, all my British friends and professors would never even notice.

So, with a jaunty step and a pink turban on my head, I took to the streets of London and headed off to Imperial College. Just as I spied my cohort of students and teachers sitting on the steps, one of them, Jack Black, a colourful fellow with an indomitable spirit, came running up to me, pointing to his own head as if to indicate mine. 'What, Kapany? Are you crazy? A pink turban? *Today?*'

Figuring he was joshing, I played along. 'Yes! It *is* a special day,' I declared. 'Very special. And I'm celebrating.'

'Celebrating?' he said. 'Even worse!' And then he paused, as if taking stock. 'Of course. Of course. You haven't heard. You, with your head full of optical glass and inventions.' He drew me closer, placing an arm around my shoulder. 'Narinder, you have to find another turban to wear today. Or just take the pink one off. King George died just a few hours ago. The country is in mourning.'

A third London snapshot I treasure is of myself and the Queen. Well, not exactly. In truth, the Queen and I never actually met. Nor does she likely recollect that I stood just a few feet away from her during her coronation procession in 1953. But I did. That was me, standing right next to the Marble Arch when the young queen in her carriage passed through, along with her husband, Prince Phillip, who, incidentally, I *will* meet. Prime Minister Churchill followed in another carriage not far behind. And on horseback, not 6 feet away from my station at the arch, rode Field Marshall Bernard Montgomery.

It was a glorious event, the coronation – a virtual Who's Who of British heroes, royalty, and statesmen. I remember looking around for someone with a camera who might snap a pic of me standing in the front row, but to no avail. Still, the memory of that coronation is as clear to me today as if it happened yesterday.

Of my early times there, I remember, in particular, a Christmas celebration dinner at the grill just off Piccadilly Square. I remember Satinder and I and a number of friends, toasting holidays past, holidays present, and any other holidays we could muster; the 'pop!' and froth of champagne bottles – I dare not venture how many or how loud we were. We abandoned the restaurant when we spied the first few flakes of snow cascading past the window to fall softly onto the square, the blurred glimmer of streetlamps projecting through their frosted lenses. Maybe eight of us in all, romping through the fresh snow, high, tipsy, and drunk with freedom and possibility.

By far, though, my most vivid memory of these long-gone London times is the day that Satinder and I married in what was back then the only gurudwara in all of Britain. Today, there are nearly 300 such Sikh houses of worship throughout the land, but none to compare in my memory bank to the one where we got married.

It was a genuine red-letter event, well attended by relatives and friends of mine and Satinder's from India and England and elsewhere, as well as by a large contingent of my professors and colleagues from Imperial

College, including the inimitable Jack Black, most senior of the graduate students, who himself had gotten married just two weeks earlier.

Following Satinder's and my wedding ceremony, we all reconvened at the Overseas League in London's West End, a tony meeting place and clubhouse housed in a magnificent historic building where Brits living overseas would congregate. Everyone at the reception was dressed to the nines, at least 100 attendees there to feast on a full dinner buffet and champagne. Neither Satinder nor I could really afford it, but somehow we scraped together enough cash to make it happen.

I'd like to think that it was a memorable night for us all. I know that it was for Satinder and me… and also for Jack Black, who was so taken by Satinder at the reception that in his toast – with his new wife sitting at his side – he proclaimed that 'Narinder Singh Kapany has taken the most beautiful woman in the world as his bride!'

Not long after our honeymoon, I ran into Jack and asked him how he was doing. 'Terrible,' he told me. His wife had evidently banned him from their nearly new marriage bed for his verbal transgression. And when he asked her for how long, she replied, 'until I say otherwise.' Smart woman.

The Sikh wedding ceremony of Satinder and Narinder in 1954, at the Shepherd's Bush Gurudwara in London, the oldest gurudwara in Europe.

Piled Higher and Deeper

It never occurred to me that finishing my PhD at Imperial College would be difficult. And it wasn't. After all, back then I knew as much about the subject of fibre optics and their application as, really, anyone in the world. Specifically, my thesis involved the design and building of a 'flexible axis optical system' that could take the place of the rigid, downright medieval endoscopes that physicians of that era used to examine your innards, and that, in at least one incarnation, would require a patient to lie flat on his back on a table with his head hanging off the side while an *in*flexible viewing tube was shoved down his throat. No kidding, it took an hour of talk therapy with a psychiatrist to ready a patient for the procedure. By contrast, with my highly flexible scope, the patient just had to swallow and the tube conformed to the contours of his throat and the other physiognomy it passed through.

By 1952, I'd already proven that flexible fibres could be used for transmitting light and images, and in the ensuing months we published three papers about the topic in distinguished journals. So, extending the technology for use in a medical/optical instrument was a relatively logical, straightforward, jump for me.

All told, getting my degree took about three years of preliminary work and research, then six months to write the thesis and prepare for my orals: a total of three-and-a-half years from start to finish.

Still, for some reason, I was more than a little anxious anticipating my orals exam. There were two examiners on my orals committee: Professor Hopkins and another professor named Tolanski, a scientist on the faculty of University of London, who was working on the development and measurement of thin coatings using interferometry. I knew what to expect from Hopkins, but didn't know much about Tolanski. Was he a wild-card interrogator? Would he throw crazy, difficult-to-anticipate questions at me?

On the day of the exam, they kept me waiting outside the examining room for almost an hour, talking about who-knows-what, before they opened the door and welcomed me in. It was Professor Tolanski who spoke first: 'Don't worry, Kapany, we weren't talking about you. In fact, I'm certain you know the stuff of your thesis so well and so much better than I, that I don't even know what to ask.'

So we talked, the three of us, for about 15 minutes after which Tolanski signed off on the exam and told me I was doing some 'lovely work,' and to 'just keep going.' And just like that, after a relatively short stint, I had a PhD.

'He'll keep going, all right,' Hopkins declared. 'In fact, we're thinking we'll be offering him a professorship.'

'Splendid,' Tolanski said as he walked out of the room. 'First rate!'

'A professorship?' I asked Hopkins. It was all news to me, the first I'd heard of it. Besides, the idea was so casually introduced, I thought that Hopkins might be joking.

'There'll be the time and the funds for research. And you'll have students, too. What could be better?'

'Returning to India,' I first thought to say. And then I *did* say it. 'Returning to India. And opening my own optics manufacturing company.' I also told him that he was the only person I'd confided my plan to other than Satinder.

'Ah, the "plan,"' he said. If he was disappointed, he hid it well.

Only a faint, fleeting frown suggested so. 'Ah,' he repeated, 'you're a good man, Kapany. And you will do what you will do. And,' he concluded, 'no doubt, it will be the right thing.'

The right thing? Did *I* even know?

The 80-Pound Car

It wasn't really that much of a car, a post-war Austin 10 that we managed to procure for 80 pounds. It was a four-door, mint-green sedan with black, tufted leather upholstery. It sounds better on the page than it looked in person. There were four of us in all: Satinder, me, and two chums of ours, both Brits from the College, off to an international optics conference in Florence, Italy, where I was to deliver my first paper to a public audience.

Our route would take us from London to Calais across the English Channel, heading for the south of France, with a two-day stopover in Paris. As I was the only one who knew how to drive, I was the sole driver.

On our fourth night in Aix-en-Provence, we parked in a dicey neighbourhood not far from the University of France, where we had been assigned a pair of miniature rooms, more down-on-your-luck than charming. I suppose I should have been more concerned about leaving my pigskin briefcase containing my presentation under the front seat, along with all our travel togs in the boot and crammed into the Austin's assorted nooks and crannies. But it was late and we were all tired and a bit snockered from our evening's libations and eager for our rooms, no matter how small. Upon departing the Austin with our toiletries, we rolled up the windows, locked the doors, and secured the boot.

Not to anyone's enormous surprise, when we arrived back at the car that next morning, all four of the doors and the boot were wide open and everything we'd stowed there the night before was gone. Everything, that is, but the pigskin case holding my forty glass presentation slides, still safe and sound under the front passenger's seat.

Fortunately, Satinder was able to catch the attention of an eagle-eyed gendarme who took us on a 10-minute walk to the police precinct where absolutely nothing was learned other than that my contingent spoke no French and theirs spoke no English. There were some documents signed, some others issued, but nothing came of any of it.

By noon, we were back in the car on the way south to Lyon, then Monaco, followed by Switzerland and finally Italy, eating, drinking (in moderation), and joking all the way. Though I was driving, I used the quiet times between treacherous mountain passes to mentally rehearse my presentation, which, for some reason, wasn't going well – not well at all. Indeed, all I could think of besides the winding road was our vandalized car.

We arrived in Florence as scheduled, a day before the conference was to begin. Professor Hopkins had already arrived and asked the four of us to join him for dinner. I did not want to worry him – fearful that a poor performance by me the next morning at the conference keynote would reflect badly on him as my sponsor – but somewhere amidst the light-as-air, meringue floating islands, I blurted out that our car had been broken into, our bags stolen, and that instead of rehearsing on our journey, I was feeling too unsettled to give the presentation its due, feeling both angry and violated.

'Ah,' he said, 'all your bags. And the pigskin briefcase, too? It was such a lovely case,' he added commiserating – how seriously, I'm not certain. Was the professor pulling my leg?

'No,' I said, 'they left the case with the presentation. Under the front seat.'

'Well, then, we're in good shape,' Hopkins assured me. 'The bounders can steal your physical belongings. But they can't take your intellectual ones, eh? Even if they had taken the case. Remember that!' he added. 'Besides, I've brought the copy you left with me in London a few weeks back, to check your facts and your conclusions. It's exactly in the same condition it was when you first gave it to me,' Hopkins admitted. 'Honestly, I never looked at it once.'

It was Hopkins's candour, cool, and humour, I'm certain, that led me to confess that I'd been quite nervous all along about the conference, it being my first public lecture, and that for my sake – and his – I had wanted it to go off without a hitch.

'Listen, Kapany, you have nothing to worry about. No one in that audience knows anything about fibre optics. In fact, you're just about the only one in the world who does. So just talk. Tell them about it. Share your excitement. That's all you have to do.'

'What about the slides?' I asked. 'Should I use those? I have about forty. Should I request a projector?'

'By all means, if it would make you feel any better. But trust me, Kapany, all they want is *you*.'

Pondering the presentation and the professor's words for much of the evening, I decided I'd do my best to arrive at the podium filled with facts and startling observations. At least the audience couldn't penalize me for poor preparation, and for lack of either data or effort. Satinder and Hopkins sat next to each other in the front row. I was slated to speak for 15 minutes. It took me two or three slides before I got my bearings and began to feel truly confident with the material again. Two or three more and I was really hitting my stride. But before I could finish the sixth, the conference moderator notified me that my time was up. Eager to comply by the rules and not overstay my welcome, I was in the process of gathering my notes and leaving the podium when I heard the first hoots from the audience.

'Let the man talk!' someone yelled with a French accent. And then someone else, an American, yelled the same, followed by a dozen more 'Let the man talk!'s, all interspersed with more hoots. It was so raucous I felt I was in Parliament. Clearly, there was a great deal of interest among the optics community in fibre optics. For an awkward moment I didn't know whether to leave or stay. It was the moderator who came to my rescue. He surrendered the stage to me, giving me a full 40 minutes to fill, which I did.

Top, left: The Kapanys with Professor Dennis Gabor, inventor of the Hologram.
Top, right: With Professor Giuliano Toraldo di Francia (far left).
Bottom: Among guests gathered at the Uffizi Gallery, Florence, Italy, for the International Commission on Optics Conference.

The Plan and the Man at the Back of the Room (1)

In the coffee break that ensued, a man at the back of the room – the director of the Institute of Optics at the University of Rochester in upstate New York, as I would soon learn – patiently waited to introduce himself, as a few final members of the audience peppered me with questions and shared their business cards with me. When finally the crowd in the front of the auditorium had dissipated, leaving only Satinder, myself, and the man from the back of the room, he identified himself as Robert Hopkins and asked if Satinder and I would join him for lunch.

It was there, in the conference venue dining room, after only a few minutes of chitchat that included a colleague who had joined us, that he said, 'I want you to come to the University of Rochester. I want you to come to America.' The table fell silent as he repeated his request. I looked over to Satinder who, not surprisingly, looked back to me. America! The word hung tacitly, pregnantly between us, and it was at least a half minute before I said, 'I'm sorry. But I can't. I already have a plan.'

Hopkins chuckled. 'Every one of you young Turks has a plan. What's yours?'

I told him my plan, the one I'd recently confided to the first Professor Hopkins in my life: to take my degree, return to India, and start my own factory manufacturing optical devices. 'That's it. My plan,' I concluded. Although, I'll admit that when I articulated it to those around the table, myself included, it didn't sound particularly exciting, or even challenging.

As if reading my mind, Hopkins said, 'Surely, Kapany, you can do better than that.'

'It's a good plan,' I insisted, looking to Satinder for affirmation. She seemed to avoid my eyes. Meanwhile, Hopkins's wingman kept insisting that we move to Rochester, trying to capture my interest by throwing out the names of nearby corporations the Institute worked with – Kodak and Bausch & Lomb, for starters – all with impressive optics credentials.

'Look,' he said, 'you're already abroad. Four thousand miles west of India. So why not stay abroad a little longer? Travel three thousand miles further west and spend a year with us. I promise you it will be worthwhile. Also, that you'll love the place.'

'Rochester, New York?'

'Right! You won't regret a day.'

I looked over at Satinder, who now batted her long lashes coquettishly at me, something she almost never did, not saying a word. So I said it for the two of us. 'America!'

Satinder and Narinder, attending the International Commission on Optics Conference, Florence, Italy, 1954.

Campari

That evening, Satinder and I attended a cocktail party hosted by one of the conference's local sponsors. It was held in the courtyard of a splendid Florentine palazzo, where our host came up to us and said to Satinder in refined, precisely accented English, 'Perhaps you have heard the comparison before, Signora Kapany…' he said, paused, then continued, 'that you look like Sophia Loren.' 'An Italian actress,' he went on to explain, 'a lovely young woman, just beginning her film career. But please,' he added, suddenly embarrassed, 'my manners are poor. What can I find for you to drink?'

I asked for Scotch, and soda and Satinder, a far more restrained drinker, asked for something simple.

'Ah, yes. Perfect,' the man said. He returned with our drinks on a small silver tray a few minutes later, 'a Scotch with soda for *Signore Professore,*' he announced, 'and for the lady, a Campari.' The liquor shone bright red in the glass, ice cubes tinkling as Satinder took it from his hand. '*Salute!*' he said, taking his own glass from the tray and raising it, acknowledging us both: 'To optical fibres!'

Satinder sipped tentatively from the glass. She even manufactured a smile to indicate her approval. And then, the second our host was out of earshot, she handed her glass to me. 'Terrible,' she said. 'Terrible, terrible, terrible. So bitter. I hate it!'

By now she had unwittingly attracted the attention of more than a few fellow guests. I took a sip of the Campari myself just to see how terrible it was. I had to agree. It was terrible, terribly bitter.

Who would've thought that, what with all those classic beach and café umbrellas that bore its name. I handed the glass back to Satinder. 'Just hold on to it,' I suggested. And so she did, my gorgeous Sophia Loren lookalike. Avoiding any further toasts.

Oh, Satinder – we had such a wonderful life together!

Mrs. Toraldo di Francia with the Kapanys at a social event during the Optics Conference, 1954.

In Delhi, 1973, with Giani Zail Singh (third from left), who became the first Sikh President of India with other Sikh leaders.

Left: With Sardar Swaran Singh, Union Cabinet Minister of India, 1973. *Right:* With Giani Gurmukh Singh Musafir, a renowned Punjabi poet and Chief Minister of Punjab, 1973.

Left: With Prof. Darshan Singh Ragi (front centre) during his visit to the Sikh Foundation office; (far right) Prof. Gopal Singh, then manager at the Sikh Foundation. *Right:* Sardar Hardit Singh Malik, Indian diplomat to Canada and France.

Left: With Dr. Manmohan Singh (the first Sikh Prime Minister of India) and his wife, Gursharan Kaur, at their residence in New Delhi. *Right:* Narinder receiving the 'Pravasi Bhartiya Award' from Atal Bihari Vajpayee (the tenth Prime Minister of India), 2004.

Left: Narinder visiting the writer Khushwant Singh at his home in New Delhi. *Right:* With economists Dr. Isher Judge Ahluwalia and Montek Singh Ahluwalia during their visit to the Sikh Foundation, 2015.

With Ambassador Navtej Sarna (front centre), visiting the Sikh Gallery at the Asian Art Museum in San Francisco with Sikh Foundation trustees and museum officials.

Narinder with Dr. V. P. Singh (former Prime Minister of India) and artist Arpana Caur at the opening of her art exhibition, New Delhi.

Left: Sardar Tarlochan Singh, Member of Parliament in India, and journalist Kuldip Nayar at Delhi during a meeting with Dr. Kapany. *Right:* With author Sardar Patwant Singh in Delhi.

Narinder receiving an honorary doctorate from General J.F.R. Jacob (far right), Governor of Punjab, at Guru Nanak Dev University, Amritsar, India, 2000.

Left: The Kapanys with actor Kabir Bedi. *Right:* (Left to right) Vice Chancellor of Guru Nanak Dev University, Narinder Kapany, Minister Murli Manohar Joshi, singer Hans Raj Hans, General J.F.R. Jacob, and other officials.

PART III

COMING TO AMERICA!

Facing page: Watercolour and pen on paper, artist Sumeet D. Aurora.

Gestation

Satinder and I sailed to New York from London in the summer of 1955 on a single-stack ocean liner of British registry that was alleged – by a very convincing fellow passenger – to have belonged to the German Navy prior to being seized by the Allies at the close of the war. Not only had she been German, our travelling companion insisted, but Hitler himself had been an occasional passenger, the Fuhrer's reputed tendency to seasickness notwithstanding.

The ship's officers and crew, mostly Brits, did little to confirm or deny the rumours, and as a consequence much of the initial talk among the passengers was of our vessel's storied past. Within a few hours of being at sea, however, the lot of us were embroiled in the pleasant camaraderie of an ocean cruise. Our vessel was fast and the accommodations were utilitarian, leaning to austere. Laughter and largely English chit-chat filled the communal rooms and decks. Hitler's ghost or not, we were determined to have a good time.

Unfortunately, the weather did not cooperate, and by our second day at sea the rain, wind, chop, and waves had become so severe they drove us all into our cabins, where we remained, save for the occasional run to the dining salon where stewards and waiters oversaw the replenishment of a never-ending buffet. Pitched and tossed about, Satinder and I filled our plates and brought our bounty back to our cabin where, by our later reckoning, our son, Raj, was conceived. Gestation was under way.

After a few days of this routine, we arrived in New York harbour with virtually every passenger standing on deck anticipating landfall, even those who fared poorly on the voyage. It was exciting to see the Statue of Liberty as we steamed past that welcoming beacon, accompanied by the bustle of small-boat traffic, the roar of marine diesels, the horns, the sirens, the klaxons, intermixed with commanding blasts from our own ship's bridge. For those waiting on the massive dock for friends and relatives to arrive – no one was expecting Satinder and me – our ship emerged from the blaze of the mid-morning sun at her stern. We must have looked almost magical, plying our way into the slip with the help of a pair of earnest tugboats, our decks filled with nattily dressed Brits, sallow-complexioned Americans, and tanned Indians, not to mention Jews and Arabs, many speaking English with different accents. Nothing, in any event, that Hitler could have envisioned in his worst nightmare.

By noon, the gangplanks were down and Satinder and I set foot in the United States for the first time, the Hudson River at our backs, the smell of brine mixing with that of exhaust fumes and, after a few hundred yards, the aromas emanating from fish, fruit and vegetable stands. I bought an orange for each of us, ripe and juicy, which we quickly peeled and ate on the spot, better tasting even than the oranges Jagdish and I stole from that maharaja's grove years ago. Finished with our oranges, we each ate a hot dog, our first ever, from a nearby Sabrett stand. America!

It was not as if we didn't have a short-term plan. We were headed to Rochester, in upstate New York, to the university and its Institute of Optics. Finishing our lunch, I looked for a pay telephone to call the director, announce our arrival, and tell him of our intention to take the early-afternoon train to Rochester from Grand Central Station, our luggage to follow.

Things did not work out as planned, however. The first time I rang the director up, there was no answer, not even after a multitude of rings. The optimistic glow on Satinder's face and the large, colourful, straw purse she clutched as she stood next to the phone booth gave me encouragement. I dialled the director a second time. Still no answer. On the third try, however, he picked up immediately, as if he'd been expecting me all along.

'Hello, Professor Hopkins,' I announced, a little too brightly, I realized. 'This is Kapany. We are here! In New York!'

'Oh, that's wonderful!' he said after a few seconds, processing, it seemed, who 'Kapany' was and what 'we' represented.

'Yes, in New York. In Manhattan borough,' I assured him, 'And we are taking the train to Rochester, New York, in an hour.'

'Oh, no, no,' he said in quick response, as if we were certainly not part of *his* plans at all. 'Not so soon.' He paused, possibly to think. 'You're in New York,' he proclaimed after a few seconds, then repeated, 'You're in New York. Have some fun. Spend a couple of days. I insist!' And before I could tell him that beyond the cash we had for the train fare we had no money but for a few Indian rupees, he hung up.

'Ah, Satinder,' I said, looking up at my lovely young wife, who looked back at me expectantly, 'we are in a bit of a fix.' And quite suddenly and rudely, New York, with its impressive skyline, its endless streams of yellow and checkered taxicabs, its cacophony of cars and truck horns, had its way with me – and I felt daunted for the first time in recent memory.

It was Satinder's perennially optimistic good cheer that quickly brought me back to myself and we decided to walk to Times Square (how far could it be?!) to find a hotel that would accommodate us for a few nights on the promise that we would pay for the room at a later time. I didn't, in any event, want to appear like a rube or a deadbeat to my prospective employer.

'I'm here in New York as a newly hired professor at the University of Rochester with my wife, Satinder, and we have arrived in Manhattan from London without money,' I explained to the front desk clerk in the first acceptable-looking hotel we walked into. More than a little uncertain of my ground, I pressed on. 'Of course, once we arrive at the university a few days from now, I could wire you the money…' I felt Satinder tug at my sleeve. In her calm, she had already noticed what I was too flustered to notice myself: The clerk had produced a key from a series of 100 or so wooden slots behind him and laid it on the counter between us.

'That sounds swell to me, professor. Welcome to America!'

In our room, our first on dry land in more than a week, Satinder and I took measure of our situation. Finding a hotel clerk who would stake us to a room for a few nights was one extremely fortuitous thing, even if only temporarily. Finding restaurants and taxicabs and tourist attractions that would also accommodate us with the promise of future payment, would be quite another. Even in America. Even with me wearing a turban, which, the desk clerk confided to Satinder when we checked out, 'did the trick' for us when we first arrived.

Fortunately, the mother of a university colleague of mine also lived in Manhattan, and when I phoned to tell her of our predicament she immediately offered to meet us at Grand Central Station with enough cash to last us a few days. Enough cash, anyway, to have the fun my new director insisted we have: taking the elevator to the top of the Empire State building, walking the caverns of Wall Street, stepping out of our hotel in the evening and into the lights of Broadway, eating at a famous New York delicatessen, and going by ferry – back into the harbour – to climb the steps inside the Statue of Liberty to view the city from her crown.

I was particularly impressed that it took only a single person to drive a transit system bus *and* sell and take passengers' tickets *and* open and close the door, whereas in London there'd be at least two people to perform the same three tasks. Clearly the Americans were an industrious lot. And, as I would subsequently discover, they were quick to forgive. Few, it seemed, held a grudge.

On our second morning in New York, I rose early to discover that Satinder had already left the room. Nervous about her being alone in the unfamiliar city, I searched downstairs, only to find her sitting by herself on the settee in the corner of the lobby. Her colourful purse was sitting on her lap like a large cat. She was dabbing at her eyes with a lace handkerchief. It was clear that she was upset. 'Why are you distressed?' I asked, sitting down next to her and putting an arm around her shoulder. 'What's so sad?'

'It's not sad,' she said, her ordinarily strong, assertive voice quavering slightly. 'It's just that everything is so big here. So big in America. It's just too much.'

I assumed then that this was merely New York City talking, the borough of Manhattan, and that she would get over it. In fact, though loath to admit it to myself back then, I felt much as she did. Still, I was resolved not to be overwhelmed by it. Like Father some twenty years ago at the Golden Temple, beckoning me to bathe with him in the sarovar, New York – indeed, America itself – was holding itself out to me, to us, and, in retrospect, to our son Raj as well, freshly *in utero*, just beginning to make his presence known.

Precisely eight months and twenty-two days later, a very pregnant Satinder – with a due date two days from then – and I were having dinner at the house of some of our new Rochester friends, Max and Edith Herzbergher, both in their fifties, and their kids. The Herzberghers were amusing and gracious entertainers, the meal was delicious, and everyone was in high spirits, especially my beautiful, expectant wife.

As the evening drew to an end and Satinder and I wrapped ourselves in our matching, fur-lined, locally purchased winter coats, Max insisted on escorting us down the brownstone front steps of the Herzberghers' row house. 'You must let me help you!'

'That's all right. No need,' I said to Max, taking Satinder's gloved right hand firmly in my left as we stood in the foyer.

'Then, I'll take the other side,' Max offered, nesting Satinder's left hand in his right and opening the door to the outside as the three of us began our descent. We weren't three steps down the stone stairs before Max lost his footing on a small patch of ice. Dropping to his knees, he partially took Satinder down with him, and it was only my firm hold on her other hand and onto the right-hand railing that prevented the three of us from tumbling to the sidewalk.

A split second later, still on the steps and a bit shaky, the three of us regained our balance. Just then, however, Satinder's water broke. Hugely apologetic for his role in her fall and absolutely no-nonsense under pressure, Max recovered his composure almost immediately and drove Satinder and me directly to Strong Memorial Hospital a few miles away. Five hours later Raj was born.

The gestation was complete.

Clockwise, from top left: The Kapanys with friends, atop the Empire State Building, New York City, 1955; Narinder voyaging from London to New York, 1955; Narinder holding his newborn son, Raj.

It was a Very Good Year

As my American professor Hopkins promised, I would never regret a day of the nearly two years we spent in Rochester. The university was extremely generous, provided us with a nice apartment, and gave me a wide variety of courses to teach. And the faculty and the students were wonderful – motivated, curious, and skilled. I also consulted on a number of projects with Bausch & Lomb as well as doctors in the medical school of John Hopkins University. Better yet, I had plenty of time left for research and did some very good work, by my own measure, that resulted in a number of papers.

When after a few months it became clear that we might stay more than a single year, the university arranged a larger apartment for us on its land adjoining the campus. Living there in that neighbourhood of single-family homes and small apartment buildings populated mostly by faculty members, we were struck by how friendly and open everyone was. Unlike living in London, where typically you didn't know anyone living in your proximity – or even in your apartment building – in our Rochester neighbourhood we came to know everyone. And everyone seemed to be throwing a party almost every night.

Which is not to say that people weren't hard-working. Quite the contrary, we found that the folks we befriended were among the most diligent we'd ever met or were to meet. During the weekdays, and into the evening, you could see kids doing chores in the garden or in an open garage. And on the weekends, the moms and dads would join in. It seemed that just about every Saturday was a spring-cleaning day with folks scrubbing barbecues, painting fences, and rototilling their yards.

The only downside: It was one of our Rochester neighbours who introduced Satinder to the marvels of modern fertilizer… but that's another story.

I still recall my British Professor Hopkins gently chiding me when I first told him of my planned move to Rochester. '*Upstate* New York? I can't believe it, Narinder. You're forsaking London to live in *upstate* New York?' he queried, pausing to let the sheer absurdity sink in, then added, 'Just white picket fences and perfectly manicured emerald green lawns.'

All true. But the University of Rochester was anything but a sleepy place, a fact borne out by five of my graduate students and me preparing papers devoted to our work together on fibre optics in just a little over a year. We presented them, the six of us, at the annual meeting of the Optical Society of America in Lake Placid, New York, where, once again, my plan to return to India, such as it was, was thwarted by a man from the back of the room.

Left: The University of Rochester, New York, whose Institute of Optics was the first in the United States dedicated to the study of optics.
Right: Satinder Kapany at the university campus.

The Plan and the Man at the Back of the Room (2)

The audience in Lake Placid in 1956 for the annual meeting of the Optical Society of America was large – indeed, enormous, by Optical Society standards. The excitement in the room where my graduate students and I presented our papers was palpable. Those who couldn't find places to sit stood for the entire three hours. When our session was over, a man emerged from the back of the room (again!) and took me aside. 'I need to speak with you, Dr. Kapany,' he said. I said 'okay' and we stepped out a side door and into the now nearly empty lobby. We sat down next to each other on a bench.

'I want you to come work for me,' he pronounced, immediately after which he told me his name – Dr. Leonard Reiffel – and handed me his business card. I read it. It revealed him to be the managing director of the Physics Department at Illinois Institute of Technology, a very heady place with a well-burnished reputation in the optics world. I looked up.

'Come work for me,' he repeated. 'You'll like what I have to offer.'

Just like the initial man from the back of the room who had made a similar offer and claim – my current American Professor Hopkins – this second man was also right. I would come to like what he had to offer: of note, heading up the Institute's Optics Department, overseeing the work of about thirty scientists and engineers. It was an enticing offer, to be sure. But I'd been here before and I knew it would take more than an offer of a great job at a prestigious institute to talk me out of my plans of returning to India and starting my own optical device manufacturing business. Not a whole lot more, though, as it turned out. There was a brief back-and-forth with Reiffel:

'You're already in America. Chicago is just an hour away by plane from upstate New York,' he said.

'But my destiny is in India...'

'Of course, it is. All I'm asking, Dr. Kapany, is that you spend a couple of years with us...'

'But Chicago is *so* cold...' I had him there. And we both knew it, though Rochester was nearly as cold, if not colder. He was silent for a moment, then went on. 'Yes,' he said, 'but you'd only be twenty-seven and the head of a department of a prestigious Institute...' He had me there.

Making the move to Chicago turned out to be one of the best business and life decisions I ever made. Working in the department with top-flight, highly motivated individuals with curious minds was totally energizing. The four years I ended up working in Chicago were the most productive in my life, with a multitude of publications and dozens of patents to my credit, a number of which Corning – a major corporate player in the optics world – was exceedingly interested in and eventually procured.

Meanwhile, as department head, I was regularly travelling to places like the Wright-Patterson Air Force Base in Ohio or the National Institutes of Health in Washington, DC, to secure funding for our program.

As an extra incentive, the Institute offered everyone in the department a cash bonus every month that he or she came up with an invention. With fibre optics still so new, everyone was inventing things willy-nilly, from ways to improve overall fibre quality to ways of magnifying or minifying fibres. It quickly got to the point for me that the notable months were those when I *didn't* score a bonus. In relatively short order, I'd become an inventing machine.

Vistas of Fiber Optics

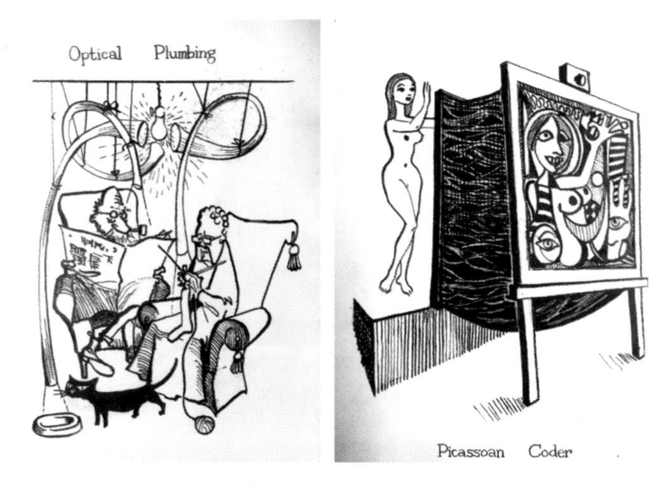

Tummyscopy

Whimsical pen and ink cartoons by Porges, 1958.

My employer also put us up in style, in a Mies van der Rohe–designed building on Chicago's South Side that the institute owned. It was a sizable, swanky apartment with lots of room for our growing family. And our fourth-floor apartment, with its largely glass façade, afforded us a stunning urban vista, with Lake Michigan in the background.

Alas, what our glass façade also offered was a view of the street below, where the homeless encamped in doorways, often aggressively panhandling during the day and lurking dangerously at night, getting into fights, snatching purses, drinking, and doing drugs. Without our even being aware that it was happening, the situation on the street was changing who we were, making us more wary and suspicious of everyone we passed. We'd come a long way from Rochester and its white picket fences.

In the summer of 1959, about three years into what I initially had agreed would be a two-year stay, Satinder and I decided to take an extended vacation. We told a travel agent that our major requirement for location was that it be rural and relaxing. More to the point, that it *not* be Chicago. She ended up booking us in a resort community in Wisconsin, just organized enough for swimming classes for Raj, and just isolated enough in our cabin that Satinder, Raj, our two-year old daughter, Kiki, and I could sit around a fire and not have to worry about what was happening on the street four stories below.

It was a great vacation, a tonic for us all, particularly the kids, then age two and three. Where they had grown nervous and tense in Chicago, after just two weeks in Wisconsin they were laughing and carefree. In fact, in just a few days' time we had all undergone a conversion. Aware of it, Satinder and I were both concerned that, returning to Chicago, we'd revert right back to our tense selves. It was that concern that convinced us to leave Chicago: to move on, to fulfill my plan, to do the right thing.

Back at work, I told Leonard Reiffel of my decision to leave and my primary reason for making it, hoping that in doing so I wouldn't insult him. 'This place, Chicago, the city… it's just not the place where Satinder and I want to bring up our kids,' I said, then added, 'besides, my plan was always to go back to India. And I think that now it's time.'

'I understand,' he said. 'I truly do.' But then he tried every trick in the book to convince me to stay: more money; a second, more prestigious title; greater research time; longer vacation time. But my mind was made up: I was going to leave Chicago and, most likely, would return to India to build my factory – my long-ignored plan. Concerned that I'd leave him in the lurch, however, I offered to stay on for a full year to give him time to find someone to take my place.

Nevertheless, while I did my best to bring my verve and excitement to work every day of that year, in my heart I was already gone, doing what I hoped was the right thing for myself and my family.

Left: The Kapanys with son Raj, Rochester, 1956. *Right:* Raj, daughter Kiki, with a little friend, Chicago, 1958.

Narinder, working in his lab at the Armour Research Foundation,
Illinois Institute of Technology, 1958.

the frontier

Published by ARMOUR RESEARCH FOUNDATION OF ILLINOIS INSTITUTE OF TECHNOLOGY

SPRING 1958

Fiber Optics—see page 18

SPIE NEWSLETTER Oct.-Nov, 1960

FIBER OPTICS

For those who missed the opportunity to meet and hear Dr. N. S. Kapany at the last SPIE Technical Symposium, we recommend his article in the November issue of the Scientific American Magazine. 10 pages of text and illustrations present the basic theory of fiber optics, suggest the capabilities and define the limitations.

Dr. Kapany is supervisor of optics research at the Armour Research Foundation of the Illinois Institute of Technology. His work in the field of fiber optics stems from a Royal Society scholarship granted him for the purpose of research.

Dr. Kapany participated in the Technical Symposium session devoted to fiber optics which was organized by Dr. Seigmund of the American Optical Company. The taped recording of that outstanding session is in the process of transcription. It will be printed in the Newsletter soon.

ELECTRONIC NEWS, MONDAY, AUGUST 5, 1957

Named Supervisor Of Armour Section

CHICAGO, Aug. 4. — Dr. Narinder S. Kapany has been named supervisor of light and optics section of the research department at the Armour Foundation here, it was made known last week.

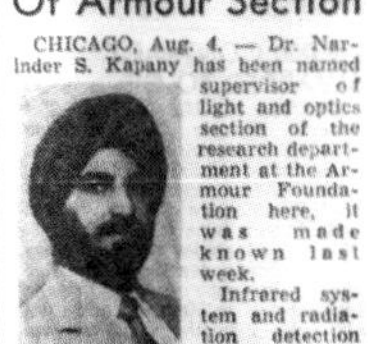

KAPANY

Infrared system and radiation detection research will come under Dr. Kapany's jurisdiction in that capacity.

12 CHICAGO DAILY NEWS, Tues., Nov. 29, '60

Next—the Doctor Will Be Peering Into Your Heart

He'll Use Bundle of Thin Glass Fibers to Look Around Corners

BY ARTHUR J. SNIDER

Daily News Science Writer

Some day a physician will be able to look inside the human heart with a kind of super-flashlight.

The device will consist of a bundle of glass fibers, each many times thinner than a human hair, but capable of transmitting light.

Because the fibers are flexible, it will be possible to bend the bundle, look around corners and see things that are hidden from direct view.

* * *

THE FIELD of science that will make this possible is called fiber optics.

Narinder S. Kapany, head of the optics research department, Armour Research Foundation, explains that the fibers are drawn from thick glass rods in a specially designed furnace. They are then coated with a layer of glass of a different kind.

If light is directed into one end of a glass fiber, it will emerge at the other end.

Bundles of such fibers can be used to conduct images over curving paths.

* * *

BUT HOW are light rays curved? Does not light travel in a straight line?

Although the rays appear to be bent, they actually are not, Kapany points out.

They follow a zig-zag path down the rod, traveling always in straight lines and caroming repeatedly off one side onto the other.

A bundle of fibers seven feet long delivers 50 per cent of the entering light at the far end. It will work effectively up to some 25 feet.

This will make it possible, Kapany adds, for physicians to examine interior parts of the body with "fiberscopes."

* * *

"THE PERISCOPE type of instrument now used for stomach examinations has several blind spots," says the optics specialist in Scientific American. "Its rigidity causes the patient considerable discomfort."

A fiber bundle will be able to reach every part of the stomach and even the small intestines beyond it and will be much easier on the patient, he adds.

"A thin fiberscope should even make it possible to scan the interior of the heart!" he added.

The Chicago American 24—Wed., July 31, 1957

Joins Tech Staff

A PIONEER in a new branch of science, Dr. Narinder S. Kapany, a native of India, has joined the staff of Armour Research Foundation at Illinois Institute of Technology. He will serve as supervisor of the light and optics section. Dr. Kapany developed a technique of using a bundle of transparent glass fibers to convey light and images along what appears to be a flexible rope of glass. This research evolved into a new branch of optics known as "fiber optics."

Press coverage about Narinder's research on fibre optics (1957–1960).

Maharaja Yadavindra Singh of Patiala (1913–1974).

The Right Thing

But what was the 'right thing'? I pondered the question daily during the year I'd given Leonard to find my replacement. Should I go back to India and start my own company as I had originally planned more than five years ago? Or should I give into a more recent impulse and move elsewhere in America – perhaps, I'd been thinking, to the San Francisco Bay Area, that high-tech mecca which, over the next decade, would become known as Silicon Valley?

I'd recently scouted out the possibility in a midwinter trip that had me taking off from Chicago's O'Hare Airport in sub-zero temperatures and arriving in the Bay Area in the middle of a summer-like 80°F with men going to their high-tech jobs in short-sleeved madras shirts and sandals. Could I possibly discover Valhalla there?

And then fate, this time in the form of a phone call rather than yet another man from the back of the room, intervened. The call was from the Maharaja of Patiala – coincidentally, the son of the very maharaja whom my cousin Jagdish and I saw passing in magnificent procession on the day we stole the maharaja's fruit. He asked if I'd come to New York City the following day, to the United Nations, to meet with him and an important friend. When I pressed him for details, he told me not to concern myself and said that more information would be forthcoming when I arrived. We agreed to meet in the Delegates Lounge.

Although I'd met the Maharaja of Patiala, Yadavindra Singh, a number of times before, I was still struck by his elegant bearing and impressive stature. He was at least 6'4" tall. He must have recognized me as well, approaching and shaking my hand the instant I walked into the room.

'It's good that you came, Narinder,' he declared. Good for him or good for me? I wasn't sure, so I asked him.

'Good for us *both*,' he replied. 'Let's sit down and I'll tell you all about it.' We found an empty table, sat, and ordered some juice when I heard a familiar, deep voice from behind me. 'Your Highness! Your Highness!' a man with a classic Indian accent called.

I didn't even have to turn to know who it was, though I *did* turn.

It was Krishna Menon, India's defence minister, the nation's second most important man at the time, and a permanent delegate to the United Nations. With a reputation for arrogance, Menon was not particularly well-liked in the States for his anti-American rhetoric and his controversial screeds in the media.

We all stood to shake hands. Menon spoke first. 'I'm pleased you came, Kapany,' he assured me. And then, without skipping a beat, added, 'Both His Highness and I think you're spending too much time abroad. Far too much. It's time you came back to India.' He paused as we sat, 'But more about that later.'

I was, of course, curious what the two men had in mind for me, but I decided to let Menon's 'later' linger over the table, and let him determine when the time was right to tell me what I needed to know. As it turned out, whatever it was wasn't sufficiently pressing for Menon to mention that day. Instead, after a perfunctory chat about my parents and my family, Menon asked that I meet him at his hotel the following morning. Then, with an obvious wink to the maharaja, he stood and walked out of the room.

'He likes you,' the maharaja said. 'He rarely likes anyone, but he seems to like *you*.'

When I arrived at Menon's stylish, old-world boutique hotel the next morning, the door to his room was ajar. Still, I pressed the doorbell. 'Narinder,' he called from somewhere in the room, 'if that's you, let yourself in. If it's anyone else, get the hell away or I'll rain all manner of security down on you.'

I accepted the invitation with some trepidation, only to see Menon emerge from what I assumed was the bathroom, dressed only in a shirt, socks, and a pair of boxer shorts. His pajama-like trousers were in one hand and in the other was the long drawstring to cinch them around his waist. 'I just don't know how to thread this damn thing in,' he said, holding the trousers and the drawstring out to me. 'You're a clever fellow, though. I'll bet you do!'

He was right. I did. I appropriated a large paperclip from a file I spied on his desk then, attaching the clip to one end of the drawstring, I threaded it through the waistband of Menon's trousers. Deeply grateful, or so he told me, he drew them on, one leg at a time.

'Listen,' he said, while we took morning tea and pastry in the sitting room, 'I meant what I said about your coming back to India. You belong there, working for your people.' He paused. 'More to the point, I've been following your career and I want you to work for *me*. In Delhi, at the Defence Department.' He paused again. 'I want you to be my scientific advisor…'

I was extremely flattered by the offer. But it wasn't in any way part of my plan, which, wherever it might take me, would, I believed with all my inventor's heart, definitely involve optical devices – inventing them, designing them, and manufacturing them. Sensing my imminent resistance to his offer, Menon asked if I'd been back to India lately. He frowned a cartoon frown when I told him 'no,' but brightened when I said that I'd planned on returning in a few months for a holiday with Satinder and Raj.

'Excellent,' he declared. 'Excellent! Just make certain you come to see me in Delhi when you are in-country. And,' he cautioned, 'don't you go making any hasty work decisions until you do. Promise?' I promised.

Six months later, leaving baby Kiki in London to be cared for by a trusted governess, Satinder, Raj, and I arrived in Delhi. The plan was to travel by car to the Kashmir Valley where we would vacation on one of the famous houseboats anchored at the Dal Lake. True to my promise, the day before we embarked on our journey, I telephoned

Krishna Menon at his office fully expecting that he wouldn't be there or have the time to see me.

Surprisingly, the phone number he gave me put me in direct touch with him rather than the first of a succession of assistants. 'I don't have the time…' he said, sounding initially grumpy, 'but for you, Narinder, I'll make some.'

'Listen,' he said when I arrived at his office less than an hour later, 'I don't need you as much today as I will in a week, when I'm hosting the Defence Service Conference. All the top military people will be there and I really want them to meet you. And you to meet them. Give a little talk. Very informal, no slides. Just off-the- cuff.'

'But I'll be with my family in Kashmir…' I began to protest.

'Of course, you have plans. Industrious, busy men always have plans,' Menon said, and then went right on. 'I'll send you a plane to fetch you wherever you are, fly you back here for two days, and fly you back to wherever you want to be. Not even enough time for your wife and children to miss you. Agreed?'

Seriously, who was I to disagree with Defence Minister Krishna Menon? So, I agreed. To what, I wasn't quite sure. Besides, it was all pretty heady business. I was not even thirty years old, and the second most important politician in India was sending a plane to fly me wherever I needed to be flown.

There was an audience of about a thousand attendees at Menon's Defence Service Conference on opening day. Highlighting that morning were opening remarks by Prime Minister Jawaharlal Nehru, a presentation by Menon himself, and another by Homi Bhabha, the head of India's Atomic Energy Commission. Fourth on the printed program was another name: 'Narinder Singh Kapany.' I could barely believe it.

Later that day, Menon took me aside. 'Mr. Nehru was very impressed with your talk,' he said. 'He wants to speak with you personally.' And, as the reality sank in of having an audience with the prime minister, no matter how brief, Menon added, 'He's cleared an hour. So, both of you should have plenty of time to have your questions answered.'

My hour with Mr. Nehru was one of the most refreshing and personally rewarding hours I have ever spent. He seemed altogether understanding and interested in what I wanted to accomplish, both in science and with my life. And not in a cursory way, but on a deeper, more emotional and more profound level. He was also very candid about his own aspirations, as a politician and as a man, and how, and whether, they had been realized. He was simply one of the most brilliant, receptive, and engaging individuals whom I'd ever met, equally able to follow the science of what I was talking about as well as its possible social and philosophical ramifications. And at some point in our quite disparate ramblings, I agreed to become Krishna Menon's science advisor.

As our talk-session was drawing to a close, and on Mr. Nehru's urging, I looked over his shoulder as he penned a note on official stationery to the Union Public Service Commission, the group in charge of the hiring of high-level Indian public servants. In it, he recommended that I be given the position of Krishna Menon's science advisor and be paid the highest possible salary for that level of government employment. Also, that I be given as much as six months to wrap up my affairs in America before starting the job in Delhi. He then signed the note, placed it in an envelope, addressed it, and laid it on top of some other envelopes on his desk. 'I should warn you, though, Narinder,' he said in closing, 'that these higher-level appointments often take time to filter through the bureaucracy – two months possibly, or even longer.'

Back on the boat in a beautiful lake in Kashmir, I shared the news with Satinder. 'I think this is really going to happen.' I recited to her, from memory, Nehru's note to the hiring commission on my behalf. 'The timing is also perfect,' I went on. 'Just about the time the new job begins, I'll be finished with my work at the institute,' I pointed out, for the first time feeling the need to convince my wife that the job plus a move to Delhi was a good idea. 'You have nothing to worry about from me,' she assured me. 'It sounds like a wonderful, if new, plan.'

'I should have an official job offer in about two months,' I said, 'though the prime minister cautioned that sometimes things take just a little longer.'

It wasn't '*the* plan,' *my* initial plan, or anything remotely like it, but we both felt thrilled by the prospect of the job. Who knew? Perhaps it was another, even better plan.

In the meanwhile, whereas before Menon's offer I was circumspect about discussing my leaving Chicago at the end of the year, I told everyone at the institute about the new plan and my direct recommendation from none other than Prime Minister Nehru. But then, as two months became three, then four, then five, and still no job offer, I began getting nervous. I called the Union Public Service Commission, and they assured me that the offer was virtually a done deal.

By the end of month five, however, with our lease about to be up on the apartment and my work at the institute drawing to a close, Satinder and I decided that no news from India was simply fate telling me that my future was not in India, after all, but in America.

So just like that, I stopped thinking about the Indian position and concentrated all my energies, sizable as they were, on Silicon Valley in sunny California. It was time, finally, to become an entrepreneur.

About a month after we settled in our new home, fully a year since the prime minister's personal recommendation, the offer from the Union Public Service Commission finally arrived.

The family during one of their vacations at the Dal Lake, Kashmir.

PART IV

IN BUSINESS

Facing page: Watercolour on paper, artist Sumeet D. Aurora.

The Bet

I'm sitting across from Merlin Dohlman at the Bank of America headquarters in the heart of San Francisco. It's 1959.

Between us is a wide expanse of mahogany desk. Behind him on a credenza are what look to be family pictures, probably his. I'm wearing a dark business suit, a black turban, and dark sunglasses, having overindulged the night before. I didn't think it then, but looking back on my meeting with Dohlman from today's perspective, I was probably the weirdest looking guy who had ever called on him. Dohlman was one of a zillion BofA vice presidents, and the most likely, by my reckoning, to give me the $500,000 seed capital I needed to fund Optics Technology, my yet-to-be leading-edge optics company. The black coffee in a porcelain cup that Dohlman's secretary brought to me spilled into the saucer. I set it down on the desk with his blessing.

'So, Dr. Kapany,' he began. His chair creaked as he moved to shorten the distance between us. 'Tell me, how much will you need to put this enterprise in business?'

'A half-million dollars,' I said, looking him in the eye through my shades and noticing a slight wince. Back in 1959, $500,000 was no small chunk of change.

With an expensive-looking fountain pen, he wrote the number down on a pad of paper, doodling small arrows around the digits as I explained to him, first, why I left Chicago (the weather, not a good environment to raise my kids); second, why I came to the Bay Area (the weather, a good environment to raise my kids); and third, what Optics Technology would, once funded, be focusing on (lasers, fibre optics, opto-electronics, and thin films), from basic research to the building, productization, and manufacture of optical devices for the military and medical markets. On that precise morning, however, all Optics Technology had was a few ideas for inventions, a team of three scientists from Chicago (including me), and a marketing specialist from the Bay Area.

As it turned out, Dohlman was a good listener, and while at one point in the afternoon confusing fibre *optics* with fibre*board* (I'm thinking now, he might have been shining me on), he seemed to get it. At least enough to know that his own bank wasn't investing in 'that sort of thing' – that is, a cutting-edge high-tech company like Optics Technology.

'But,' he went on, 'say we *were* a half-million-dollar partner in your company, what per cent would you be willing to give us?' He was leaning closer now, elbows on the desk like a schoolboy.

'A minority. A minority share,' I pronounced, without giving it more than a second's thought. 'You see,' I went on – to a bank VP, no less – 'you see, I just don't trust the money people.' I continued to look him in the eye, all the while thinking, *Oops! Maybe I've gone too far in my damning of the money people.*

Just then, though, a rich, deep laugh erupted from way down in Dohlman's gut. 'I love it,' he said, "a minority share." "Don't trust the money people." You are a wise man, Dr. Kapany,' he pronounced. 'And because you are, I'm going to send you to a venture capital firm who may just cut you the deal you're looking for.'

Both Bill Draper and Fred Anderson of Draper, Gaither & Anderson, an early Silicon Valley venture capital firm, were retired Air Force generals, with Anderson having led the air attack on Berlin during World War II. Both men were also early US ambassadors to NATO, so they knew something of the real world and the potential use, military and otherwise, for fibre optics–based technologies. In fact, Wright-Patterson Air Force Base in Ohio, one of Draper's prior commands, was already in negotiation with me to develop an infrared fibre optic device – once, that is, I had a company that could manufacture one.

It was to these two men, and about a half-dozen others in the firm, that I presented my business plan. Included in the meeting were a few high-level executives from Hewlett-Packard. They were interested in what I was up to and would become involved in the early stages of my company's funding, if not directly then indirectly.

After some preliminaries in the conference room, I reconvened alone with the two generals in Anderson's office. It was he who spoke first: 'You can stop fretting, Dr. Kapany. Provided we can work out the dollars, we're going to do this thing.' And then he posed a question: Did I know a Mr. Malik? Anderson mined his memory for any specifics, aware that the relatively common Indian name 'Malik' alone might not suffice. 'A Mr. Hardit Singh Malik,' he soon went on. 'He lived in Delhi.' Anderson smiled, warming to the memory. 'The ambassador to France! And a helluva good golfer.'

'A family friend,' I said, pleased that he had taken me to familiar territory. 'Whenever our family travelled to Delhi, we would stay in his home. He is also a Sikh,' I added, a factoid that was met by a brief silence by both generals.

And then, seemingly out of nowhere, Anderson popped the question: 'How much?' Could it be that Dohlman hadn't told him the dollar amount I was looking for? In any event, unlike the BofA's Dohlman, he didn't wince when I said 'a half million.' But he *did* laugh, as Dohlman had, when I insisted I'd only offer a minority stake in the company. No technologist, Anderson was still a thoughtful, savvy businessman, and was able to discern more than a glimmer of the enormous potential offered by my team's optical technologies.

'I'll tell you what we're willing to do,' Anderson said after a brief and knowing glance at Draper. 'Our proposal is this: You and your team – call them what you like – will own 49 per cent of the company; we, the investors, will hold another 49 per cent of the company.

'And the remaining 2 per cent will stay in a voting trust that will only revert to you when you show your first $100,000 profit in a calendar year.'

After a brief silence, Draper added, 'Don't get us wrong, it's fine with us that it be your company and that you and your team are the majority shareholders. We want it to be yours for as long as you can make a success of it. But if you can't, we'll want to be able to take control and get things back on track.'

It wasn't exactly what I'd come for, yet it was close enough. After a few more remarks by Anderson about Mr. Malik's golf game, we shook hands all around and Optics Technology was born – just eight years after first demonstrating fibre optics to my astounded professor in London.

A thick, typed, boilerplate-laced agreement followed by the end of the afternoon, but not before General Draper and I would shake hands to seal another less formal agreement about money. 'Oh incidentally, Narinder,' Draper said as the meeting was about to adjourn, 'how long do you think it will take you to find your first customer?' Based on our discussions, he knew it would be a government contract I would be angling for. And with his extensive stint in the military, he knew how long the traditional procurement process took.

What he didn't know was that I already had a contract super-loaded and shovel-ready at none other than Wright-Patterson Air Force Base, where Draper had once been C.O. All I needed was the cash that Draper, Gaither & Anderson was now making available to fulfill my end of the deal. So I told Draper I'd have my first contract signed, sealed, and delivered within six months, confident that it would likely happen far sooner than that.

'I'll bet you a hundred dollars you won't have a contract in six months,' Draper insisted. He extended his hand for me to shake on the wager.

'Better watch what you're doing there, Bill,' Anderson cautioned, chuckling, 'Dr. Kapany might just have a card up his sleeve.'

'He may,' Draper said, 'but it'll still take him more than six months to play it.'

'You're on!' I said to the general, giving his hand a vigorous shake.

When, contract in hand, it came time to collect on my bet a few months later, Draper made a show of going through his wallet and only finding a rumpled $10 bill. I let him off the hook for the other $90, provided he'd sign the ten-spot. He did, gladly I think, and General William Draper's $10 still hangs framed in my office.

Nowadays, getting a contract with the government requires jumping through all sorts of hoops. By contrast, I remember back in the 1960s and '70s, during the Vietnam War, if I came up with a new idea for a better night-vision device, for example, all I'd have to do to get funding would be to talk to the C.O. at Fort Belvoir in DC, tell him how much money I needed, estimate how long it would take to build, and I'd have a check in a month or two, or three at the most. Back then, it was just that simple.

Top: General William Draper's autographed $10 note.
Bottom: Narinder at a business meeting.

Optics Technology

It was interesting stuff. Exciting, actually, our work at Optics Technology – coming up with fibre optic devices like endoscopes to look at the insides of everything from nuclear reactors to the human heart. We were shedding new light, enabling new measurements in the widest variety of environments. One day we came up with a laser for eye surgery, and then on another we developed the world's first fibre optic card reader.

There was hardly a month that passed without a news story in the *Palo Alto Times* about the company and the work we were doing. We were going gangbusters. And we were growing – from fifteen employees when we opened our doors to four times that by the end of our first year.

Also in that year, not long after we were funded, Bill Draper called with a proposition. Draper, Gaither & Anderson had invested in a certain company, called Spectracoat, located a few miles north of Palo Alto in a town called Belmont. 'Their president is a nice old guy' – 'old' then meaning about fifty, in view of the youthful minds energizing Silicon Valley – 'but his company just isn't going anywhere. They're stalled,' Draper said, stopping to mime 'stalled,' then went on: 'I told him about you and suggested that his company and yours would be a good mix.' Draper paused again, this time for the idea to sink in. 'You'd be in charge, of course. It'd still be your company –

Optics Technology would, that is. You'd just be taking over his. Move right in.'

It sounded fairly straightforward, but still, a takeover, even with everyone in agreement, could be fraught with human resource issues: of note, a mass of disconsolate, unhappy, and angry people. As it turned out, my concerns were unfounded. The 'old guy' was dying to get out of the hot seat and was pleased that I named him director of Optics Technology's coating operation – essentially what he was doing at Spectracoat, but without having to run a company. Better yet, his people were excited by what Optics Technology had in the pipeline and were eager to be a part of it.

Also, at about that time, I approached the National Institute of Health in Maryland with the rather unique proposal that they help fund the company, something that the NIH had never done for a for-profit organization. In my six-page letter to them I made the case that modern optical technologies, such as lasers, fibre optics, and other related optical tools, would soon be mission-critical to the medical community, and that their support would keep the country ahead of the curve in this key area of basic research and product development. To my pleasant surprise, the NIH agreed to underwrite us to the tune of $250,000 a year for a number of years.

A few halcyon years into this arrangement, Ig Lou, one of our engineers, brought me a laser he had built – a neodymium-doped glass rod about three inches long and one-quarter inch in diameter. I turned the laser in my hand as he told me his plan to 'lase' a variety of human tissues to see how they would respond and interact with the laser beam. I told him I thought it was a terrific idea but that he should give some thought to taking it a few serious steps further. Ig, appreciative of the support, warmed quickly to my suggestion of using the laser to repair a detached retina: first, by affixing the laser to the end of an ophthalmoscope; second, by using this device to search the eye for the errant retina; and third, by firing the laser into the lens to reattach it – 'retinal coagulation,' as it would eventually be called. If Ig's and my process worked as I posited that day, complex retinal repairs could be achieved without costly surgery.

There was no mistaking Ig's excitement to my suggestion: 'I can do that,' he said.

He did. And it worked like a charm. First, we tried the laser itself in our preliminary tests on raw flesh – not human flesh (that would come later) but on a piece of raw beef, a steak from a local market. In short order, we felt sufficiently confident in the procedure to test the device on eyes: first on the eyes of animal cadavers, then on the eyes of live animals, and then, finally, on human cadavers. In each case, we'd focus the laser on the retina, fire the beam, and almost always register a successful repair.

After a few months of animal trials, we were eager to test the technology on humans. Early in 1961, having successfully fulfilled the regulations set forth by a then more tech-friendly Food and Drug Administration, we contacted two noted local ophthalmologists – Chris Zweng and Milton Flocks – at the Stanford Medical School to discuss the possibility of their using the device in their practice, or at least testing it further. They were excited both by the technology and by its efficacy, based on the animal experiments they conducted in their own labs, essentially repeating and reaffirming our preliminary results.

Our first human trial of the device was performed on a scientist from a local think tank (then called the Stanford Research Institute, now SRI International), where he, himself, was involved in laser research. While playing tennis, a ball had hit him on the head and the impact detached his retina. At first, he was highly resistant to our use of the technology to cure his ailment. 'No way,' he insisted, 'I'm using lasers to cut bricks! And you want to use one on me? In my *eye*?' He ended up coming to visit me at Optics Technology where we talked at length before he consented to the procedure.

Our second operation was performed on no less notable a personage than Bob Hope. Both Hope's and the tennis playing scientist's procedures were completed successfully. Since those days, laser retinal coagulation has been used on millions of patients around the world, and today, a half-century later, it continues to be standard procedure.

As for Optics Technology, while we may have been a bit cavalier about testing our product at the outset, as the company moved forward in the field of laser retinal coagulation, we became quite conservative. On the suggestion of Tom Perkins, a sales engineer and our eventual marketing director, we developed working relationships with two more luminary ophthalmologists, one in San Francisco and one in Boston, providing each with a retinal photo coagulator. In return, we asked these two men to recommend other highly credible physicians/researchers who might be interested in the device, and who would provide detailed reports on those patients for which the device was used. It was our hope in this way to build an increasingly sizable user database.

This go-slow-but-steady approach worked, with 40 devices sold to our original base of physicians/researchers at somewhere between $8,000 and $10,000 per unit, and with the per-unit cost to us of $2,500. All told, we sold hundreds of units during my tenure at the company.

Another Optics Technology–originated idea, this time cardiac-based, also resulted in a highly successful product that is still widely used today. This *in vivo* cardiac oximeter used fibre optics to measure oxygen saturation and other critical cardiac values, to assist surgeons in planning therapeutic procedures.

Also developed at Optics Technology was the helium–neon (HeNe) laser – the first application-specific laser of any sort on the market. Working on the project, I envisioned a wide variety of engineering uses for the technology, including as a means of measuring the power output, lifespan, and reliability of a product. Consequently, when it came time to put the laser on the market, I recommended a relatively high price, $600, to our board of directors – though it only cost about $100 to make. But Tom Perkins wasn't quite as excited by the technology as I was. He saw its value more in the educational realm than as an engineering tool, and suggested we price the device at $200, the upper limit affordable to most teachers in the 1960s.

I suppose I could have gone to the wall for my $600 price, but I didn't think it was worth the disharmony it might create in the company. As a side note, I was also unaware that Perkins had *already* gone to the board behind my back to have me kicked upstairs as chairman and have himself installed as president, in charge of day-to-day operations.

Internal politics notwithstanding, we settled on pricing the device at $200. To be sure, more than a few teachers became customers. Perkins was right on that account. But also becoming customers were just about all engineers in the entire country! Applying 20/20 hindsight, we would have had multiple millions of dollars more to fund other projects and to make an even bigger difference in the productivity of engineers around the world, if only we'd priced the HeNe laser at the $600 price point I originally envisioned.

Aside from its technologies and applications products enabling healthier lives for patients, better tools for engineers, and safer outcomes for the military, Optics Technology also became a boon to our investors. Founded in 1960, the company went public in 1967 at $12 a share and reached $60 a share within six months, peaking at $73 a share in 1968 and generating plenty of happy campers. By 1973, however, after thirteen years as president, I felt it was time for me to move on, even though the company was still my baby, and hard to let go of.

Besides, as I realized back then, what I really liked – what I really *loved* – was coming up with a new idea or a new thing, building it, commercializing it, and then moving on to the next thing. But when that 'thing,' whatever it was, became huge and required a large number of people to manage it, I would lose interest. As a result, and throughout my adult life, I was always looking for – I always *needed* – something fresh to excite me.

In the early 1970s – even before I left Optics Technology – that something new was… sculpture.

Science of Fiber Optics Is Led by Indian Expert

Attempt to cut gap between development and use of optical tools is aided by small company

MEDICINE IN THE MAKING

[illegible]

[illegible]

New medical tool looks inside you

PALO ALTO TIMES, WEDNESDAY [illegible]

[illegible]

Light follows curved path

[illegible]

Top, left and right: Press items about new endoscope ('light follows curved path') and fibre optics. *Middle:* Opening day of Narinder's Optics Technology company, Stanford Industrial Park, Palo Alto, California, 1964. *Bottom, left and right:* Narinder with colleagues at a meeting of the National Inventors Council, Washington, DC, 1970.

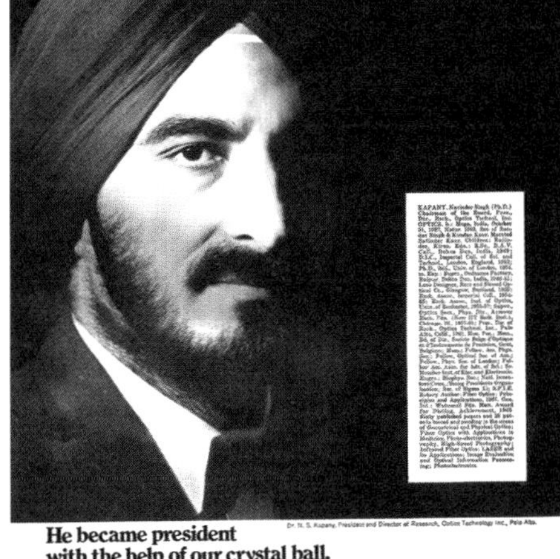

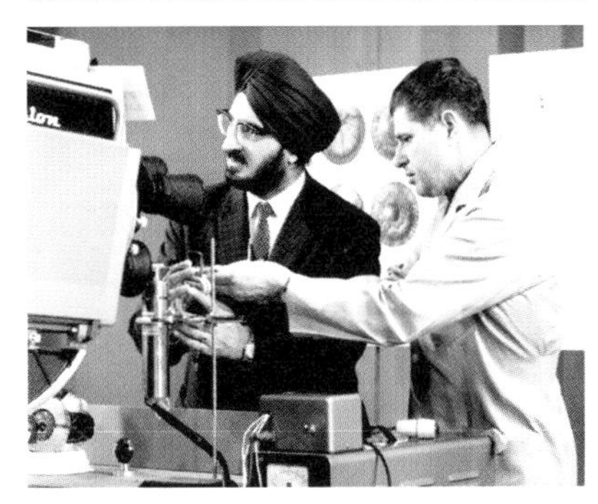

Clockwise, from top left: A fibre optics demonstration at Syntex Corp.; Narinder signing a contract with Syntex Corp. leaders; Interview with officials at Lockheed, 1960; Advertisement for Bank of America; First TV interview in San Francisco; With Professor Arthur L. Schawlow, co-inventor of the laser and a Nobel Laureate.

Searching for Something New/Veni, Vidi, da Vinci

Other than my father, my greatest hero – as a young man in the 1970s, and today – has always been Leonardo da Vinci. It's a kinship that goes all the way back to the time when Father first asked, 'What do you want to do?'

I recall telling him as a youth that I wanted to go to college, learn how to make a living. 'Yes, but aside from that…?' he persisted. I said I might go on for a master's degree. 'Yes,' he acknowledged, 'but aside from *that*?'

He wanted to know, of course, what every concerned parent wants to know: what his son would *become* in the world, what I would *do*. What label society would affix to me? Scientist? Inventor? Physician? But even back then I balked at being categorized. I'd also recently read a biography of Leonardo, and wanted something of his rich, multifaceted life for myself. I wanted to excel, not only in one thing but in many. I believed – and I hoped not naïvely so – that, given the opportunity and the resources and the curiosity, individuals were infinitely capable of wondrous things, each in his or her own way.

As for *me*, even back then I was interested in how things worked and how they were made. I was interested in light and heat and politics and munitions and even a painting of a Sikh maharaja that hung in a favoured spot in my father's ancestral home. I was interested in family and property, wealth and want, even need and poverty. I was interested in everything.

So it was hardly a surprise – to myself and to those who knew me – that, as CEO of Optics Technology, I would be a curious boss. It was a curiosity that manifested itself in the early 1970s by my walking from cubicle to cubicle, desk to desk, and lab to lab in our snug working environment at Stanford Industrial Park to ask everyone what they were working on, the challenges they were facing, the obstacles standing in their way.

It was on one such prowl that I made a remarkable find in the wastebasket next to a fibre optic extrusion machine – a device, incidentally, that I had invented in the 1950s to make optical glass fibres, that was still in use back then in the 1970s and continues to be used to this day. With it, you can heat a high-quality, pure glass rod within a glass cylinder of lower refractive index and extrude the resulting fibre onto a drum. And there in the wastebasket that evening were the remains of a failed extrusion with the fibre from the rod bent over and curled in on itself. I leaned over to pick it up. It weighed about a pound and took up roughly a cubic foot of space. I turned it this way and that, finding it pleasing to the eye. *Very* pleasing. 'This could be a sculpture,' I thought.

So I took it home to my workshop; built a simple, elegant, black box; added some lights to illuminate it; and placed the malformed extrusion in the middle. I then surrounded it with a series of narrow, vertical pillars of glass and gave it a name *The Caged Serpent*. It took me about a week of precious spare time (I was still running a company!) to complete. And the whole time I was working on it, ideas for other sculptures kept popping into my head.

They did not go ignored. Over the next two years I created about sixty such sculptures, utilizing a wide variety of materials, including fibre, white light, lasers, and interference filters. Some of these pieces were small and compact, like *The Caged Serpent*. Others were large and sprawling, like *Metropolis*, which evoked a grand city transforming from dusk to dawn, the night light ever-changing as it was captured and reflected off the surfaces of the city's spiral glass skyscrapers and lower-lying landmarks, also glass. Resting on a scrubbed brass base, the sculpture incorporated numerous lights placed behind rotating interference filter wheels and Lucite cylinders of various diameters and lengths. The light was transmitted through the cylinders, producing an ever-changing glow as the 'evening' drew on.

Over the two years I created the bulk of my sculptures, I worked in complete obscurity, which was fine with me. I loved the work and, most of all, I enjoyed exercising that part of my imagination that did not have to be productized or monetized. I was happy simply creating – for the lack of a better word – *art*!

So it came as some surprise to receive a phone call one day from Frank Oppenheimer, the founder and director of the San Francisco Exploratorium, a huge, sprawling, iconoclastic, largely hands-on science and art museum and sometime-gallery, all housed back then in a wing of the Palace of Fine Arts, a highly distinctive structure built as part of the 1915 Panama-Pacific Exhibition.

'Kapany?' He didn't wait for a response. 'Frank Oppenheimer here,' he said by way of introduction. 'I hear you've done some sculpting.'

'How did you know?' The only Oppenheimer I'd ever heard of was Robert, the 'Father of the Atom Bomb,' who, as it turned out, was Frank's brother.

He said he wanted to see the pieces. 'That's really the next step,' he pronounced. 'Actually, the first step, eh, Kapany?'

He sounded like a good fellow… and turned out to be. Strange, peculiar even, but a very good fellow. Or so I sensed over the phone that first time. So, I invited him to our house for dinner. After all, he wanted to see the pieces; and that's where the pieces were. Fortunately, ours was a good-sized house, large enough to accommodate my hobby. Leonardo Da Vinci's manifold hobbies – maybe not. But mine, yes. 'And,' not wanting to be rude, I concluded, 'and please bring your wife' – in case there was one.

Strange, peculiar, and *prompt*, Oppenheimer, accompanied by his wife, arrived right on time. Satinder was in the kitchen when I opened the front door for them and invited them in. I shook his hand, and as I did, he barely acknowledged me. Instead, he walked right past me into the living room, where he began what would become his self-guided tour through the entire house. He did this all with the support of a cane, as I recall, whose tapping was interspersed by his heavy footsteps.

By then, Satinder had arrived in the foyer and the three of us – Satinder, Mrs. Oppenheimer, and I – exchanged niceties as I heard Oppenheimer opening doors and climbing up and down stairs. For some reason, he seemed to know the sculptures were scattered through the house, as well as in my downstairs workshop. I turned away from Mrs. Oppenheimer to catch up with my guest when his wife touched my arm. 'It's okay,' she assured me, 'he'll be back in just a moment. I'm certain.'

Satinder glanced quickly at me. Between the two of us, we tacitly decided to indulge his eccentricity. True to Mrs. Oppenheimer's prediction, her husband, a tall, balding man, reappeared in the living room about 5 minutes after his abrupt departure. The three of us were ensconced in the living room as Oppenheimer joined his wife and Satinder on the couch, at which point he produced a flat, worn, leather notebook from his pocket.

'Let's see, now,' he said thumbing through the thin volume. 'The da Vinci show ends in February 1973; I believe that's the right date.' He glanced around the room as if checking with us. Not finding any disagreement, he continued, '...and after that we'll have an exhibit of the Kapany sculptures.' He wrote something in his notebook and then slapped it closed. 'Dinner?'

'But wait,' I said, fearing that by agreeing to a show I was suddenly asking to be taken seriously as a sculptor, which would no doubt involve being lambasted by critics as an upstart, an amateur, a no-talent businessman/entrepreneur/inventor, a dabbler, a poseur. 'Wait!' I repeated. 'This is something I do for fun. A hobby. It's just very simple.'

'No,' Oppenheimer replied, refusing to brook my demurral. 'No, I think this "hobby" of yours is something strong. Maybe ten years from now when everyone is doing it, doing fibre-optic art-science that is, your *dynoptic*' – he seemed to have coined the term on the spot – 'sculptures will seem simple. But right now, *today*, they are *new*.'

The San Francisco installation of my dynoptic sculptures was complex and required my being on site at the Exploratorium much of the time to ensure that everything was working as it was designed to. As for the critical reviews, there weren't many, but those that did appear were good, leading to more exhibitions at Stanford University, Syntex Pharmaceuticals, museums in Monterey and Carmel (both in California), and a major museum exhibition in Chicago, which involved packaging up the whole kit and caboodle for a multi-day journey in each direction.

Best of all, though, for my own psyche if nothing else, was the timing of the first show, the one at the Exploratorium where the works of Narinder Singh Kapany from Moga, India, followed right on the heels of the works of none other than Leonardo da Vinci.

Thurs., June 1, 1972 ☆ S.F. Examiner—Page 33

Fascinating Art of Glass and Laser Beams

By William Zakariasen

"One should never ignore the esthetics of science," says Dr. Narinder Kapany, president of Optics Technology, and whose new exhibit of optical sculpture is enchanting visitors to the Exploratorium in the Palace of Fine Arts.

India-born, London-educated, Dr. Kapany has been president of Optics for 11 years, and during that time he has developed several inventions, including a device for making color X-ray plates. He has also expanded the possibilities of laser beam techniques for use in medicine. However, he is a Renaissance man, deeply interested in the arts as well.

A little over two years ago, Kapany, through a fortunate accident, was able to make his dream of some sort of artistic contribution come true. In the field of optics, there is much waste of glass and fibers — most of it is naturally destined for the trash bin. One day, the remains of a fibre parent rod (the material used to protect tubing) caught his eye. It was in the form of a graceful snake. He expanded it to his first art work, "Caged Serpent."

Fascinated by the possibilities of this new art form, Kapany has continued his designs from these waste materials into an almost psychedelic series of visual confections. The multi-hued collection now numbers several dozen, and the artist-doctor is constantly adding to it. "I'm a weekend artist, but there's a lot of thinking beforehand," he adds.

A slow tour around the circular room housing the Kapany exhibit is like being inside a kaleidoscope. Forms and colors are constantly changing, and several pieces will never look the same again. One is titled "Ecstasy," and it first seems merely a series of colored squares. Yet, through clever use of interference filters, the colors change as one walks by — most eye-catching.

"Tranquility," a favorite of many visitors, is a series of gray and white panels composed of neutral density filters, and "Ballet Africain" is a madcap of tiny laser mirrors. "Night Sky" is a huge inverted bowl, inside of which is a slowly rotating crystal, reflecting the white light and laser beams.

"Habitat," inspired by the edifice of the same name at the Montreal Expo '67, is an engaging series of blocks housing several tiny laser beams, almost invisible to the naked eye, and "Agaricus" is described by the artist as a depiction of "cosmic mushrooms."

"Metropolis" is a special favorite of the artist. Again, it is a seemingly nondescript pile of tubing, but one which changes into excitement through the play of laser beams on the interior kaleidoscope.

Other intriguing titles, at times self-explanatory or at least provocative, include "Pleafull Prayer," "Space Ecology," "Solar Panel" and "New Beacon." Dr. Kapany's turbaned, bearded presence at his exhibit, moreover, lends an aura of sorcery to the titillating sights.

The exhibit is scheduled to end June 15, though its popularity, particularly with the young, may keep it in the Exploratorium somewhat longer. "Already, several other museums have approached me, saying they will gladly take over the exhibit when it's through here," he says.

"Nevertheless, I enjoy showing it here in San Francisco," he adds. "I often come up from my Palo Alto office in the late afternoon before closing, just to watch the reactions of children and young people to my work. It's a very satisfying experience."

Just then, a group of grade-schoolers "oohed" and "aaahed" in the presence of "Metropolis," and the omniscient face of Dr. Narinder Kapany broke into a broad smile.

DR. NARINDER KAPANY WITH "METROPOLIS" AT THE EXPLORATORIUM

Waste materials from his optics company became visual confections

—Examiner photo by Bob Bryant

Press report about Narinder's own fibre optic artworks, in the *San Francisco 'Examiner,'* 1972.

Tempo

Family Focus — Section B — FRIDAY, APRIL 5, 1974 — Garden — Page 13

'Aquatown'

Narinder Kapany and 'Metropolis'

Futuristic sculpture exhibit in Palo Alto

A "futuristic" new style of sculpture that uses precision glass, lasers, fiber optics, colored lights and scientific techniques will go on exhibit Monday at the Syntex Art Gallery in Palo Alto.

These optical sculptures are the creation of Narinder S. Kapany, a widely known physicist, inventor and entrepreneur. He is generally regarded as the "father of fiber optics."

It was a chance observation of the serpentine shape of a fiber optics rod in a trash bin that sparked Kapany's imagination with the possibili... using such materials...

Since the creation of "Caged Serpent" in 1970, Kapany has come up with many more complex and sophisticated works using his unique blend of precision glass, light and physics.

One of the largest of these in the Syntex exhibit is "Night Sky." At first it appears as simply a large ..., but on closer inspection the viewer ... note the simulation of ... and clouds ...

Kapany, who is president and chief executive officer of his own optics firm, Kaptron Inc. of Palo Alto, ... his unusual art pieces ... sculpture."

... use of special filters. ... forms and colors also can be seen when the sculptures are viewed from different angles.

Kapany's sculptures have been exhibited previously at the Chicago Museum of Science and Industry and the Exploratorium at the Palace of Fine Arts in San Francisco. This will be the first showing of his works on the Peninsula. After the Syntex show closes on May 10, th... several other ... and overseas.

The India-bo... ed Kapany ha... sued and appl... optics. As an a... nearly 80 scie... books, both of ... among the mo... fiber optics.

Kapany has ... for 13 years, a... own company ... director of rese... ogy Inc., also ... Bernard He... ounts the ... bed Kapany ... the aesth... "a gracef... or expres...

"Caged Serpent" (1970), one of the works in the Syntex show.

Prophet of techno-art shows work at Syntex

By C.E. MAVES

There are any number of art shows on the Peninsula, but the most mind-bogging is doubtless the Narinder Kapany exhibit at the Syntex Gallery, 3401 Hillview, Palo Alto.

Kapany, a physicist widely known as "the father of fiber optics," calls the 22 works in the exhibit "dynoptic sculptures." Others may well call them "freaky" and "psychedelic." They are certainly unusual, for instead of marble and bronze they employ such materials as precision glass and laser beams.

These are the creations of a scientist on holiday, of a man for whom art is an imaginative extension of everyday practical interests.

Some of the earlier sculptures, dating from 1970, re... conventional mosa... "Africain" and ...

Much more compelling pieces from this period are "Agaricus" and "Solar Panel." They are bright, hard-edged confections of plastic, machinery sublimated into art. They seem taken from the control room of the spaceship in "2001," hinting at functions the viewer can barely imagine. Playful, even mocking, above all else cool, they dazzle without drawing you in.

"Space Ecology," however, another sculpture from 1970, implies mysteries not merely technical but spiritual as well. Its phosphorescent glow and serenely floating shapes are also evident in "Monument," a chameleon structure that appears lime green when viewed from the side, and hot pink from above.

Kapany's best work, then, is characterized by futuristic evocations and elusive ... what is evoked ...

work, work overtime, and make permanent impressions. Two works stand out as the show's high points. One of them is "Habitat," a collection of transparent glass cubes inhabited only by odd glancing flashes of red light. The light is produced by lasers, as the viewers discovers when he holds his hand over the object and himself becomes inhabited.

Even finer, perhaps, is "Metropolis," for which the delightful "Aquatown" seems to be an earlier and smaller study.

"Metropolis" is a series of cylindrical linked modules, glowing pastel green and blue from within and washed in red light from without. Its expertise is akin to the tension, grace, and fantasy of the most advanced jazz. It suggests both the fragility of a dream and the inevitabilities of the future.

The Narinder Kapany exhibit will continue through May 10, and anyone who ... see tomorrow ... a point ...

MUSEUM OF SCIENCE & INDUSTRY

SPECIAL EXHIBITS AND PROGRAMS

JULY/AUGUST/SEPTEMBER

Clockwise, from top left: Press reports; Brochure of the Museum of Science & Industry; Professor Frank Oppenheimer, director of San Francisco's Exploratorium; One of Narinder's 'dynoptic' sculptures, titled *The Caged Serpent*, 1970; dynoptic sculpture, *Metropolis*, 1971.

Delhi Again?

Back in 1967, when I took Optics Technology public, I was among the first Indians in America to take *any* company public. It was an accomplishment that, as it turned out, accorded me some attention both in the United States and back home in India.

Come 1973, when I sold my share of the company, my colleagues at the Young Presidents Organization (YPO) – where I was the sole Indian – suggested that I might make Washington, DC, my next stop, notably as the Assistant Secretary of Commerce in Science and Technology, in what turned out to be the twilight of President Nixon's second administration. I was flattered by their support, but dubious about even going for an interview. The truth was, I'd never liked Nixon, never trusted him in all the years he was in political office. Still, I thought that working in public service in Washington for a few years might be a welcome change from my entrepreneurial ways.

So it was with some trepidation that I asked the management of the YPO to submit my request for an interview, only to learn that there would be *two* interviews: the first with the President's appointments secretary to get on Nixon's calendar, and the second, possibly days or weeks later, with the man himself.

Nixon's appointments secretary seemed nice enough, an affable, dark-haired fellow from Southern California. We sat across from each other on cushioned, hard-backed chairs in a small, oak-panelled office.

He had a comfortable, soft-spoken manner that immediately put me at ease, and I was glad that I had made the trip. Then, halfway through my response to his question about my personal goals, the fellow leaned way forward, placing his hands on his knees and said, 'That's all very good, Dr. Kapany, but how do you actually *feel* about the president?'

Somewhat surprised, if not completely ambushed by the question, I looked at him quizzically. He leaned forward even closer and looked me right in the eye as if we were about to exchange the deepest of intimacies. 'Nixon,' he pronounced, 'what do you think of Nixon?'

I tried to avoid the question by telling him something like 'I never mix my personal politics with my work,' but he had me there. He'd read my dislike for the man. We both knew it. And that was the end of that. And not long after, that was the end of the Nixon Presidency.

A second opportunity for a possible posting in government in 1973 was as the ostensible second-in-command to the incoming US ambassador to India, Daniel P. Moynihan. It was Luis Alvarez, a winner of the Nobel Prize in physics and a member of the Optics Technology board, who made the initial contact. 'He's a good friend,' Luis said of the ambassador-to-be. 'You should go back to India with him. Move to Delhi!'

Leave California to start a new life with Satinder and the kids in Delhi? If it hadn't been Luis who suggested it, I wouldn't have even given it a second thought. Besides, I felt that that ship had already sailed a few years back, with the bungled offer from Krishna Menon and Prime Minister Nehru.

On the other hand, working for the State Department in an interesting and challenging posting had genuine appeal – particularly if it would allow me to help build a new and deeper understanding between the United Sates and India. That would be the real challenge. I was a bridge-builder back then, and continue to be today. More than ever, in fact. So, I decided to meet with Mr. Moynihan to hear what he had to say.

He had yet to assume his post in Delhi and was still living in Boston at the time we first met. In the cab from Logan Airport to his house in Cambridge, I was more than a little excited. It was winter in Boston, overcast and snowing, as we passed the familiar brick and white-columned façades at Harvard. I had come prepared for the trip in a heavy overcoat with a fur collar that last saw action some years before when we were still living in Chicago.

Mr. Moynihan was not at home when I arrived, but his wife invited me in, dusted off my overcoat in the foyer with a flourish, and sat me down in the living room. 'The weather,' she offered by way of apology for her husband's absence, 'It slows everything down.' I recalled the summers in India and concurred. Snow can slow you down, but heat can drop you in your tracks, I remembered.

Mr. Moynihan arrived some minutes later. We had a long, leisurely lunch. I don't remember what we ate, but do remember helping to finish a bottle of wine. My host and I discussed the current Indian political scene and what a posting to Delhi for an American-born statesman like himself might mean. And also how effective – or possibly how *in*effective – an ambassadorial tandem with an Indian expat like myself in the number two position might be in 1970s India, a considerably less enlightened and less cosmopolitan India than today's.

Still, we parted on a positive, optimistic note in the mutually voiced hope that together we might be able to 'make something click', as Mr. Moynihan put it. It was not to be the case, however, and largely, I believe, for the very trepidations he expressed in our conversation that day: that it was simply too early in modern India's development for a partnership such as ours.

Mon., August 20, 1973 DAILY INDEPENDENT-HERALD—

India ambassador visits

T.N. Kaul (right), ambassador to the United States from India, was a guest of the local East Indian community Saturday. Among those having lunch with the ambassador at the Uriz Hotel were (from left) Consul General S.K. Bhutani, Mrs. N.S. Kapany, wife of the ambassador's host; Sutter County District Attorney Dave Teja and Dr. Kapany.

Narinder and Satinder with India's ambassador to the United States, T.N. Kaul (first from right) and other officials, 1973.

Narinder, boarding a flight to Europe, 1963.

'If There's a Fork in the Road, Take It'

In late 1973, with neither the position in Washington nor the posting to Delhi working out, I had the leisure to explore the world for other provocative possibilities.

It wasn't long before I found one such possibility and fixed on it. And then, in a few years' time, I found another, a *second* possibility, and fixed on *that*, too. That the two demanded totally different disciplines, were located 40 miles apart and separated by a mountain range, and were both ostensibly 'full-time,' didn't make either less appealing. I loved them both. Until, eventually, I had to let one go.

I gave birth to the first of these two possibilities in the back seat of a taxicab in Paris where I'd just spent two days consulting – under the aegis of a tiny, one-horse research company I'd founded called Kaptron – with a major French telecommunications firm that was doing some exploratory work in fibre optics. They'd flown me to the capital to share what I thought about the possible use of fibre technology in the telecom industry. I'd spoken and listened to their R&D group and a few marketing types, offering suggestions, before I turned their question back on them:

'When you perfect the part that you can already do yourself, what are you still going to need?'

My question was initially met with silence, but then the most voluble of their people offered, 'You've got light, a signal coming in, you want to be able to split it. You'll want it to be able to carry multiple channels, switch channels. You'll want to multiplex it, send video links through it.'

'All *right*!' I held up my hand. 'Enough! Everything you say makes sense to me. Let me have a few weeks with it. I'll see what I can come up with.'

As it turned out, all I really needed was the 45-minute cab ride from the International Hotel to the Charles de Gaulle Airport to arrive at a solution for splitting a light-borne signal into multiple channels. By simply placing the fibre in front of a spherical mirror at a distance twice the mirror's central curvature, you'd be able to put that identical image on the mirror's opposite side without any loss. And in between you could do anything you wanted to the light from the fibre. Switch it. Move it. Put different wavelengths in it.

Back in my lab in California, I noodled some more with the idea to get a better sense of precisely how my spherical mirror fibre-optic device might be designed and what it actually could (or couldn't) do. I also took my idea to Fred Unterleitner, a scientist with some theoretical and experimental savvy who had worked with me over the years.

'I've got an invention,' I said to him. 'I want you to figure out (a) how we can build it, and (b) *if* we build it, what its value might be.'

Meanwhile, I started to make a few phone calls of my own – to Bell Labs and other telecommunications companies. Each one said that an optical device capable of such things would be very valuable to them. Better yet, that moving forward, they would need lots of these optical devices. Fred concurred.

And so it was that I turned my small, inconsequential research company into a thriving telecommunications device designer and manufacturer. Initially located directly across the street from Optics Technology, Kaptron would become home to dozens of products for which we secured a number of patents. It was also a company that I would sell twice, and, depending on how you count it, almost a third time.

And thereby hangs another tale.

The Little Company that Could... and Did

Just about the time that fibre optics became a go-to technology for a host of high-tech applications, the telecommunications industry was becoming a major fibre user. And everyone in both industries was looking for a bigger piece of the action, as were totally unrelated industries looking to get into high tech in any way that made sense to their directors and investors, no matter how far the reach.

This was precisely the situation when Kaptron was contacted by the Crown Life Insurance Company, a Canadian firm with a hankering to get into both high tech and fibre. They'd already opened their coffers and acquired a number of high-tech companies in Boston and elsewhere, and came to us, cash in hand – *all* cash! – with a very generous, straightforward offer. The only hitch was that they wanted me to stay on to run Kaptron – this, when I'd grown tired of directing a product company and was looking for other ways to spend my time. To make the deal that much more appealing, they even set us up in new and larger digs, a 40,000 square-foot building less than 5 minutes away from my old Optics Technology office. It was an offer I *couldn't* refuse. At least, I *didn't*.

To my surprise, and that of the entire Kaptron team, less than two years after the company's purchase a pair of Crown Life executives dropped in on Kaptron for lunch. Unfortunately, they explained over a second glass of very good wine, Crown Life's decision to go into high tech wasn't panning out and they were in the process of pulling out of the industry.

'Completely,' the more insistent of my lunch companions stressed, as if Crown Life's venture into a foreign land was just a bad trip, one that he was more than ready to be done with.

'So, what would you like us to do with your company?' the less insistent executive asked. Good question, especially since the company wasn't really mine any more. I was only running it. Still, moving forward, I knew that for Kaptron to continue to remain competitive would require a new infusion of capital that Crown Life was clearly not about to make. And I, knowing the market and still excited about the products in the pipeline, was more than willing to take the risk. So I offered to take their own company off their hands, leaving them with a 10 per cent share, no matter how much I put into it. Not surprisingly, they took the offer. And, as simple as that, we owned 90 per cent of a company that we'd been paid millions for just two years ago.

It wasn't long after that, that little old Kaptron captured the attention of Corning Industries in my former stomping grounds in upstate New York. A deal, in fact, for the acquisition of Kaptron by Corning was thought by both parties to be imminent. All that stood in the way of the transaction's going through was a simple up or down vote by Corning's board of directors.

Facing this last obstacle – I was assured a 'yes' vote would be pro forma – I experienced a moment or two of trepidation. First, that the deal might *not* go through: After all, the millions that Corning was offering for Kaptron was a lot of money, even if I'd sold the company once before. My second concern was that the deal might actually go through. Like Optics Technology before it, Kaptron had been my baby since its conception in the back seat of a Parisian taxicab. I was, in other words, just a little more than reluctant to give it up.

So it was with rueful acceptance that I weathered the surprise announcement out of Corning that its board had nixed the offer. Not the entire board, only the faction that followed the lead of none other than Tom Perkins, who, as it turned out, was now a Corning director as well as a successful venture capitalist.

I suppose I should have done my own due diligence and scoured the names of the directors as soon as Corning came calling; or, saving that, if I only could rewrite history and not promote Perkins at Optics Technology in the early 1960s to vice president, a title he clearly coveted but that I denied him when I witnessed him chewing out the head of engineering in front of his co-workers. But all that was water over the dam, more than a decade before the Corning non-acquisition, and I doubt it would have made much difference in that 1970s board meeting.

Besides, it wasn't long after the Corning deal collapsed that we were contacted by another large company, AMP Inc., that was eager to acquire Kaptron. Without a whiff of Tom Perkins in the air, the deal was done. I sold the same company twice.

Better yet, this time there was no hitch. I didn't have to run the company. Instead, AMP's VP, Jay Hassan, offered to make me an AMP Fellow with the responsibility of running their research activities around the country, though principally in Pennsylvania, working with their team of top-flight engineers and scientists, inventing my own stuff, helping others run their projects, and, in some cases, expanding them. It was all great fun, a job that had been dropped in my lap, one that I performed and enjoyed for several years.

The AMP position also provided me with the satisfaction of working more as a teacher and mentor than as a boss, and, I'd like to think, doing it well. Not long ago, a former AMP scientist I worked with dropped in my office for a drink and told me that I was the best manager he'd ever had. I was both surprised and flattered. 'Why was that?' I asked. 'Because,' he went on, 'you always seemed to understand what we loved about our work and were respectful of it. And you'd let us do it without butting in or insisting that you do it *for* us.'

I briefly thought about what he'd just said and then added, 'I suppose it's because I'd done it all myself at some point: Had an idea. Brought it to fruition. Even built my own instruments. Virtually every one by hand. Also, by the time I got to AMP, I was older, more mature, and felt less of a need to compete with others in the field.'

Advertisement for Optics Technology, Inc.; Narinder, president and director of research.

SOME OF NARINDER'S SCIENTIFIC INVENTIONS

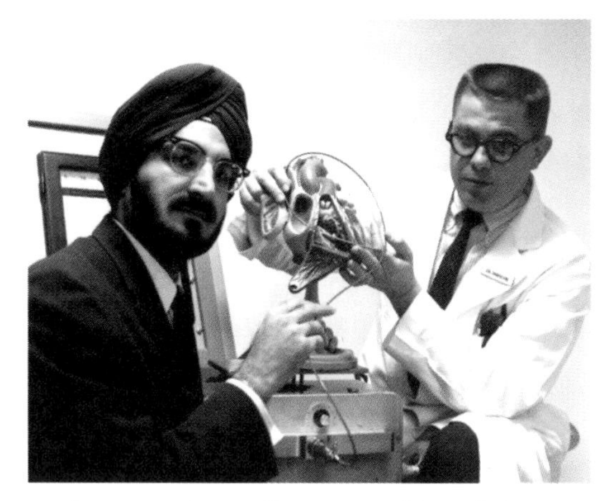

Demonstrating an Oximeter used in cardiology.

Laser Coagulator.

Demonstrating fibre optics during a TV interview.

SOME OF NARINDER'S SCIENTIFIC INVENTIONS

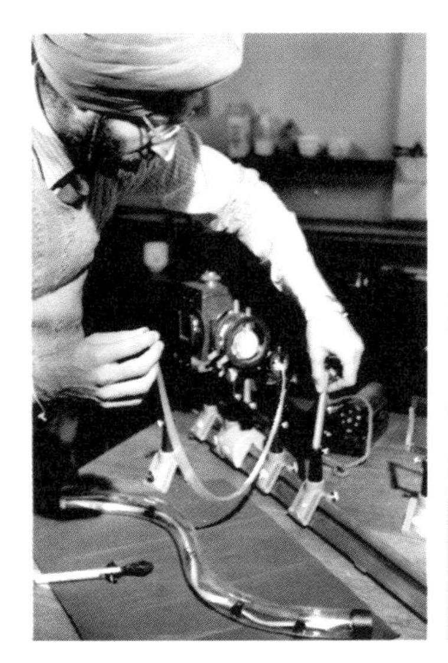

Left: Working on a bundle of aligned fibres at Imperial College London. *Right:* Conducting tests on fiber optic bundles.

Testing the endoscope with colleague Hal Sowers.

PART V

TEACHING

'Anything You Want'

The second possibility that captured my imagination, won my heart, and consumed hundreds of hours of my time in the 1970s – well before I'd sold Kaptron a second time and while I was still guiding the company through a heady period of its growth and development – was teaching.

It was an interest that was sparked in the 1960s by a pair of brief but plum teaching assignments in the physics departments both at the University of California, Berkeley and at Stanford, where I discovered that I loved the give-and-take of the classroom and the lab. I particularly enjoyed interacting with young students excited about their ideas. Not only did I see myself in these individuals as I had been some ten to twenty years earlier, I saw the potential for a whole new generation of thinkers and doers.

So, a decade later, in 1975, when I was invited to spend six months at the University of California, Santa Cruz (UCSC) as a Regents Professor, an appointment most often given to individuals who bridge the academic and non-academic worlds, I didn't waver. Nor was I concerned for a moment that teaching would steal energy from my work at Kaptron. Rather, I believed that teaching would be the perfect complement to my work, no matter how time-consuming or how frequent the treacherous 40-mile drive over the Santa Cruz mountains from my home or office to the university. Besides, as a successful inventor, entrepreneur, and businessman, I'd already burnished my non-academic credentials. It was now time now to build my teaching résumé, slim as it was.

Early on in my tenure as Regents Professor, I devoted most of my effort to giving public lectures at the university and at civic venues in and around the small city of Santa Cruz, where I spoke to faculty and students as well as local residents – usually about innovation, the productization of ideas, creativity, productivity, and (always an audience favourite) entrepreneurship. For whatever reason, whether it was ingrained in the American heritage or a product of the times and place – that is, the Silicon Valley – people just seemed to love learning about what went on behind the making of something from nothing.

I began to attract some attention. With my turban, long beard, and deep voice, I was certainly no low-profile appointment. And my talks – which appeared to those in the audience to be given fearlessly, but actually had me roiling and nervous in my gut every time I'd give them, even just a few words, and especially on my favourite and most beloved topics – tended to be quirky and provocative, often more out-of-the-box than in.

Not surprisingly, one individual in particular who took notice was the man who'd hired me, the university's chancellor, Robert Sinsheimer, himself a scientist researching the human genome. We began that year as colleagues with some mutual interests, and by the end of my Regent's appointment, we were good friends. Nearing my last days on the campus, I dropped by Bob's office to say goodbye.

'Goodbye?' He seemed honestly surprised. 'Yes,' I said, 'my six months. They're up.'

'Nonsense,' he said, then added, 'you don't think we're just going to let you go, do you? No way!'

Friends or not, I was surprised by his candour and his apparent certainty that I would stay. Like my father years before, and others since, he seemed to know more about me, at least in this instance, than I did myself. 'You're not going anywhere, Narinder.'

I was flattered, of course, and said as much. But seriously, both he and I knew there was no real place for me at the university.

'Listen, Bob,' I began, feeling the need to state the obvious. 'I *really* appreciate your interest. But I'm basically a physicist, an inventor, an engineer, an entrepreneurial type. And you don't have an engineering school here. Nor do you have a business school. What would I *do*? Where would I fit in?'

Bob absorbed this, then slowly began, 'You could,' he said, articulating the next four words precisely, most likely for impact, 'do… anything… you… want.'

Anything I want? Anything I want! Was this yet another offer I couldn't refuse? It sure seemed like it. My initial impulse was to simply jump at the offer. But instead I kept my head, tailoring the measured tone of my response to that of his offer: 'Give me a month,' I said with equal precision, 'and I'll come back to you with a proposal.'

'And whatever it is,' Bob assured me, 'we'll try to make it happen.'

Narinder, teaching a class at University of California, Santa Cruz.

Making it Happen/Seeds

'Entrepreneurship is ingrained in the American heritage and system of free enterprise.'
– Narinder Singh Kapany (from my 1981 testimony to the US House of Representatives Subcommittee on Science, Research, and Technology)

Undaunted, eager, and on schedule, precisely one month to the day of telling Bob that I'd be back, I returned to his office. And, as promised, I came with a plan. It was as simple and complex as this: to create a comprehensive, reality-based and -driven educational program for men and women seeking to become entrepreneurs. A place, in other words, where UCSC students could bring their ideas, inventions, and innovations – or could generate new ones – turn them into reality, and then build a viable business around them.

It didn't take Bob long to respond. Only a few seconds. 'That's great, Narinder.' He stood to shake my hand. 'Terrific.' And before the meeting was over, working together, we gave the programme its name: The Center for Innovation and Entrepreneurial Development, or CIED (pronounced 'seed').

As much as CIED was our brainchild, we couldn't have implemented it without the help of my fellow physicist and UCSC faculty member Professor Bruce Rosenblum. It was Bruce who helped me anchor my ideas for the centre in the real world, specifically as they involved dealing with the UCSC faculty, something I was definitely not an expert in. Bruce somehow knew precisely when to share our ideas with the faculty, and when to isolate me from my academic colleagues. And how to do both in an elegant, tactful way – keeping in mind that while today a program like the one we envisioned would hardly raise any eyebrows, back then it was downright revolutionary. The only other school that had a programme like CIED – focused exclusively on innovation and entrepreneurship – was MIT.

In researching our plan, it actually came as quite a surprise to discover that American business schools hadn't yet picked up on the need to develop structured programme to help guide those starting their own companies. Instead, they were more geared to preparing students to hold down mid-management positions in large corporations. Not even the Stanford Graduate School of Business – in the very heart of Silicon Valley, and the incubator for many of America's top-rung high-tech companies – was seizing on the opportunity to provide for this new generation of business pioneers, high-tech and otherwise, that was emerging. Many, incidentally, with non-business, non-engineering or non-tech backgrounds.

In fact – and in retrospect – it was the unique, diverse, multidisciplinary, and sometimes eccentric makeup of the UCSC student body as a whole, as well as that of the individual students signing up for what appeared innocuously on the 1980 fall quarter class schedule as 'Economics 16, Innovation and Entrepreneurship for the 1980s,' that made the three quarters CIED sequence (Economics 16, 17, and 18) such an instant success. Indeed, even in the program's early going, the approximately 150 students that signed up for 'Econ 16' hailed from a wide variety of majors, including English, philosophy, the physical sciences, and others.

Well and good. The problem with this enthusiastic response, however, was that Bruce and I had determined on an optimum class size of 30. With five times that number signing up, I decided to 'qualify' students for the course by having them fill out a questionnaire we prepared. One question from the dozen or so we asked: 'What would you do if you made a lot of money?' The right answer, by our lights? 'Donate some to charity.'

While somewhat effective, this winnowing process still left about 50 students whom I tried gently to discourage from taking the class by citing the inevitable downsides of being an entrepreneur: never having as much money as you need; never having enough time to get things done; and, a hidden factor and outgrowth of the other two, the virtually certain sowing of disharmony among family and friends.

Of course, any student hell-bent on becoming an entrepreneur would not likely be discouraged by something as inconsequential as giving the 'wrong' answer on a professor's questionnaire or not getting into an undergraduate college course. Doggedness is an important attribute in a field where, by most measures, only a fraction succeed.

As for those who remained, following my questionnaire and other discouraging tactics, they were a ragtag but highly motivated group of thirty students eager for an introductory quarter that provided the promised introduction to entrepreneurship: creating a business plan; protecting an innovation and turning it into a business endeavour; and organizing a venture. The final classes in the quarter featured visits by a number of successful contemporary innovators and entrepreneurs.

The highlight of the second course in the series, Economics 17, was the jewel in the crown of the programme, as far as I was concerned. Here, students got to work with a cadre of luminaries whom Bruce and I had assembled from a variety of fields to provide guest lectures and labs in their individual areas of expertise.

Included on this list were patent and intellectual property attorneys, engineers, venture capitalists, economists, corporate board members and directors, marketing researchers, finance and banking experts, and of course, entrepreneurs – people who had founded companies large and small.

Not only did the participation of these individuals enable a unique interaction with the business and industrial communities for students, the involvement of these heavy hitters coming from San Francisco and

Silicon Valley to Santa Cruz leant a high degree of credibility to the program as a whole.

As the programme evolved, working with these guest speakers constituted about 30 per cent of the second-quarter course – called simply 'Entrepreneurship' – to include participation in laboratory sessions, where students worked in teams to develop their own projects. Some 70 per cent of the remainder consisted of lectures that covered subjects such as market research, financial planning, and the organization of a venture.

Come spring, students following the course sequence would enroll in Econ 18, 'New Venture Planning.' In this course, class members worked in teams of five or six, focusing on one specific innovation or new product, with the goal of creating a comprehensive and realistic business plan that could, in turn, result in a viable commercial endeavour. Students were also given real-life titles (President, Director of Research, Marketing Director, and so on), and, in accordance with their titles, assumed real-life responsibilities.

As in Econ 17, students evolving these plans would frequently work in close collaboration with outside inventors and entrepreneurs, including guest lecturers, essentially bringing the real world into the classroom and occasionally taking what they learned in the classroom out to local businesses.

Also very real about the CIED programme – initially funded through the UCSC Chancellor's discretionary fund, and subsequently by the National Science Foundation – was the significant number of minorities and women in the program, more than a few of them going on to found companies or becoming entrepreneurial employees in other organizations.

Yet the number of those who eventually made money as entrepreneurs and still continue to work in the business is limited. That's the nature of the beast. Of the dozens of students I taught through the Center over six years, fewer than ten became truly successful as entrepreneurs.

One notable success is Dan Pulcrano, who purchased a number of local newspapers and created a mini newspaper empire, publishing as many as five papers at one time. Dan has since sold his first company and started a second multi-newspaper publishing enterprise that is doing extremely well, even in the face of the diminishing interest in and declining readership of print media. Still, despite the soft marketplace, I feel optimistic about his company's prospects, as both a mentor and an investor.

Another success story built on ideas nurtured by the CIED program is that of David Silverglade, who founded a toy company and subsequently sold it at a substantial profit. As did Dan, David has since started another successful company where, once again, I'm pleased to count myself as a satisfied shareholder.

A third CIED success story involved an already existing company, Southwall, headquartered in Palo Alto, which was in the business of designing and installing specially coated custom windows. The company was floundering when I accepted the offer to sit on its board. The president and founder was a perfectly nice fellow, but not particularly entrepreneurial. Nor did he have a good grasp of what was happening in his market. All he knew for a fact was that his local and international investors were losing patience.

So, I made Southwall's board an offer: 'Let me put one of my CIED teams on the job and I guarantee we'll come up with something to save the company and satisfy your investors – a Plan B, if you will.' Which is precisely what the team did. Working through the summer, five CIED graduates and I developed an alternative to Southwall's business model and strategy.

The board approved the idea, but didn't think that the existing president was the man to implement the revised programme. So, they let him go and hired a new president, who worked with the team to make our Plan B a success. Not long after, the company went public.

My seven years as director of the Center were some of the most gratifying of my professional life. UCSC had a reputation as a counterculture school, particularly in the late 1960s and '70s, and was a magnet for undergraduates open to experimentation, innovation, and highly original and creative thinking. It's hard for me to imagine a school and a student body more suited to a program like CIED. Working on original ideas in a woodsy, rough-and-tumble environment under a canopy of redwoods and eucalyptus trees was, in many ways, the exact opposite of the classroom and lab experience other students had in the more traditional business and professional schools where, quite frankly, original ideas were not at the core of the curriculum.

As for me, working in that unique environment, I gained as much as a person and an innovator as I gave. It was only the runaway success of Kaptron and my ultimate inability, my hubris notwithstanding, to both manage that company and do the job that I felt needed to be done at CIED, that had me resigning from the Center. Still, I have only the fondest memories. And somewhere in my basement, I have sixty or seventy business plans prepared by my students.

Keeping those times alive for me are the frequent emails and letters I still receive from former students, as well as my occasional get-togethers with them, sometimes resulting in interesting revelations. Not long ago, for example, I was home sipping the first of my two evening glasses of wine when the phone rang. It was one of my favourite students from some ten years back, and I recognized his voice immediately. It turned out that he and four others from his CIED class were having beers at a bar in Santa Cruz and they wanted to know if I could join them. 'We have something to ask you,' he said somewhat mysteriously. It was quite a drive to Santa Cruz from my home in Woodside, but Satinder was out of town, and I figured, why not? Besides, what question, after an entire decade, would they have to ask?

I arrived at the bar to a jovial greeting, a sudsy toast, and, before 5 minutes elapsed, the question. It was the ringleader who posed it, the fellow who'd made the call to coax me from my quiet comfort at home to this rowdy bar in Santa Cruz. I'd made the trip over the mountains so many times, I fancied I could do it blindfolded, though this was hardly recommended.

'Dr. Kapany,' he began, lending me the title many, many years after it was no longer the expected address, 'Dr. Kapany, we all know you've done many wonderful things,' the ringleader began as the six of us settled in comfortably around a table toward the rear of the room, where it smelled of spilled beer and peanut husks. 'But tell us,' he went on, 'even you, you must have made some mistakes. What are they? Or if you made only one, what was it?'

'Aha! So that's the question!' Simple as that, I thought, recalling that I'd wondered similarly about my past mentors – about Professor Hopkins, in particular – though I'd never dared to ask. As it was, ours was a relationship that didn't turn out as well as I might have hoped. The professor's support for me in my initial research into fibre optics was of inestimable worth, which made our parting on less than an altogether positive note all the more disturbing.

Our disagreement, as so many are, was about money – specifically, about patent rights. Rather than hold equal rights to our discoveries, Professor Hopkins proposed that I hold the rights in India and that he hold them for Europe and the United States. That didn't make any sense. Besides, what ever happened to Professor Hopkins's avowed Communism? Ultimately, we agreed to split everything equally, but not until we'd had a number of uncomfortable discussions.

But enough about the more distant past, and back to Santa Cruz, my ex-students, and the question at hand.

'Of course, I made mistakes,' I began, as the six of us huddled closer around the table like conspirators, as if to prevent any confidential information from escaping from our midst. 'Of course,' I repeated, my tone mock-confessional as I began…

'Back when I started my first company, I was surprised that, along with it, I'd suddenly acquired an inordinate number of friends. They'd invite me to parties and sporting events, leave me tickets for shows, ask me to join them for lunch and often dinner.'

I paused, then went on. 'At first, I chalked up all this newfound camaraderie and friendship to living in California. And to the fact that the Bay Area was a lot less formal than the East Coast or the Midwest or England or India, where I'd lived previously. And then one day, when I was attending an American professional football game thanks to the largesse of one of my 'friends,' I experienced a blinding flash of the obvious: I realized that all these people were my *friends* because I was in the position to make them money. And the instant I was no longer able to do so, they took their 'friendship' – along with their tickets – with them. Which is to say, they weren't my friends at all. Never were. Mistake number one!'

The six of us grew quiet. It was the ringleader who broke the silence. 'If you'll allow it, Dr. Kapany, *we'll* be your friends. Along with you being our mentor. Well, you're *already* that.'

'Of course, of course,' I said, accepting their offer graciously and even gratefully. I was genuinely flattered. 'A wise choice,' I joked. 'It might even stand you well,' I went on, recalling my relationship with a fellow named Alex Balkanski, whom I had also served as mentor. 'It might even stand us *both* well.'

'Pretty cryptic,' the ringleader chuckled. 'Sounds like you have a story to tell us.'

'You're right about that,' I said, and proceeded with my story about Alex, starting at the very beginning, on the day he and I *didn't* meet.

'Alex's father, Minko Balkanski,' I began, 'was a university physics professor teaching in Paris. Independently and fabulously wealthy, he and his wife lived in a most luxurious apartment with full-on views of the Eiffel Tower. It was obvious that Minko was proud of the place, and on the first day we visited him there, he insisted on giving Satinder and me a tour of the apartment. We followed close behind him as he threw open door after door, revealing at least a half-dozen splendidly decorated bedrooms, a stunning chef's kitchen and pantry, various brass-fixtured-and-trimmed bathrooms, a host of commodious cedar-lined closets, and an elegant living and dining room. Only one door went unopened, and Minko, Satinder, and I stood before it. "That's Alex's room," Minko proclaimed quietly, reverentially. "Our son. He sleeps late."'

The six of us around the table in the Santa Cruz bar laughed. 'Alex was about ten then,' I said, 'and as I recall, I didn't meet him that day. But I did meet him countless times after that, most often at the family's country home, a genuine palace about 60 miles from Paris, which the Balkanskis had bought and refurbished.

'No spoiled brat, as his father's reverence toward him might suggest, Alex was an extremely bright and curious kid, wise beyond his years and with a terrific sense of humour. And for some reason, he loved hearing my stories. All of them, some of them again and again, especially those of my naughty boyhood and of my business ventures. So it came as no surprise, really, that some fifteen years later, after Alex finished his PhD at Harvard, started a company of his own, and moved to Palo Alto, he called me up.

'He was still looking for a proper place to set up shop, so I rented him and his fledgling company half the building that Kaptron was housed in, about 20,000 square feet, and we became neighbours. More than neighbours, friends; or, even more accurately, mentor and mentee. He'd come to me with questions nearly every day – about business, about life – and I would do my best to advise him. I remember, for example, talking with him at length about the problems and conflicts he was having with his company's president, Alex being the company's Chief Technical Officer. I told him of my similar experiences with difficult-to-get-along-with C-level employees, of note Tom Perkins, though back then I did not mention his name.

'Owing, Alex claimed, to our near-daily discussions and my thoughtful counselling, he managed to convince his nemesis to move on a few months later. Alex subsequently took the company public and made a good chunk of money: enough to gift me 3,000 shares and eventually leave the company to become a successful venture capitalist.'

'Oh, I see,' the ringleader joked. 'You were just in it with Alex for the shares. And with *us*, too, no doubt!'

'Hardly,' I said. 'I'm in it for the camaraderie, for the friendship, for the reminder of wonderful times gone by.'

We were toasting the sentiment when suddenly, amid the conviviality and good fellowship, a dark shadow streaked across my mind. It caught me completely off guard. What's this about, I wondered? And then, just like that, in the midst of our celebration of good times and my affection for Alex, I remembered the worst mistake I'd ever made in my working life, the one I'd conveniently, to this moment, forgotten. And here in a Santa Cruz bar redolent of suds and peanuts I felt I had no choice but to reveal it.

'Of course,' I began, my voice hoarse, 'I've made many a mistake over the years. But only one truly critical one… and it shames me no small amount to confess it.'

'You don't have to,' the ringleader assured. 'Not at all. Not another word.'

'No. No,' I insisted, and went on. 'When I was a young man, younger than you are now, I was invited to give a series of lectures in upstate New York to an audience of even younger students, most of whom were just beginning to think about subjects that I'd been researching for a decade. So, of course, it would stand to reason that they didn't have the depth of knowledge or understanding that I did. Of course not,' I said quietly, more to myself than my audience, my voice breaking slightly. 'Nor did I exactly win them over,' I went on, 'when one or the other of them made some ill-informed or downright stupid statements. Or asked a foolish question.'

I paused, then launched into what turned out to be my mea culpa. 'Instead, I was stupid and foolish myself. I lorded it over them, saying things like, 'You have no idea what you're talking about,' sometimes even saying these things in public and in front of their peers.' Indeed, much to my chagrin, I had done to them roughly the same thing I'd taken Tom Perkins to task for.

I paused again. 'It was a terrible mistake,' I proclaimed, both judge and jury. 'A terrible, terrible mistake. Here I was, a supposedly well-brought-up Indian gentleman and a well-educated Brit, one or both. And yet, I was dismissive and arrogant. And, more than likely, I made some serious enemies. Probably even more than a few. Lifelong.' I slumped back in my chair. Having confessed this to my hail-fellow-well-met ex-students, I felt both embarrassed and contrite. 'A terrible, terrible mistake,' I concluded.

And not even the earnest, jolly, welcoming ringleader was able to put my mind at ease.

A few days back, I received an email from my old alma mater, Imperial College in London. The college was, the email said, developing a program on entrepreneurship. And, as an exemplar in the field, would I consider sending them some information about my past successes. And please, it was noted, the sooner the better.

Ah, the British, I thought, though generously and with great affection and appreciation. Two weeks after Brexit and thirty-five years after the inception of CIED, someone was mounting an entrepreneurship program at Imperial College! Sooner *would* be better. A lot better. So, I wrote back and told the writer of the letter about CIED and what I'd been up to all these years, prompting the near-immediate response…

'Help!'

Clockwise, from top left: Narinder with Dr. Alex Balkanski at his wedding in France; with students David Silverglade and Dan Pulcrano from University of California, Santa Cruz.

LEADERS *from the* WEST

Left: Narinder greeting Prince Philip in London. *Right:* Meeting US President Bill Clinton.

Left: With the Hon. Harjit Singh Sajjan, Canada's Minister of National Defence, during his visit to the Sikh Foundation, 2016. *Right:* Satinder and Narinder with First Lady Hillary Clinton at the White House.

Left: Chancellor George Blumenthall and Narinder with Dr. Sue Carter, the Narinder Kapany Professor in Entrepreneurship, University of California, Santa Cruz, 2017. *Right:* With Chancellor Henry T. Yang of the University of California, Santa Barbara.

Left: Mrs. Chester Bowles and Mrs. G. J. Watumull present Narinder with the first Distinguished Achievement Award of the Watumull Foundation. *Right:* Narinder with Yogi Harbhajan Singh Khalsa at Española, New Mexico, in the 1970s.

Left: With San Francisco Mayor Willie Brown Jr. during the Sikh Foundation's 35th Anniversary celebration at the Asian Art Museum, 2003. *Right:* Lord Indarjit Singh of Wimbledon; Sonia Dhami, Executive Director of the Sikh Foundation; and Narinder, 2013.

Left: With Dr. Mohammad Qayoumi (seated, centre), President of California State University, East Bay. *Right:* Interview with filmmaker Gurinder Chadha at an event at the Victoria & Albert Museum, London, 2009.

PART VI

FARMING

A Sikh, a Farmer

What first brought me to London is easy enough to explain: a curious and talented professor and the opportunity to earn a PhD. What brought me to Rochester: excellent researchers and facilities, plus a crack at working with some of the top companies in my field. To Chicago: my own optics department and the time and support to do great work. To San Francisco and the Silicon Valley: an entrepreneur's Valhalla.

What brought me to Yuba City and Fresno, California, however, was a whole other story, one that had absolutely nothing to do with optics: farming.

It was a story that, for Sikhs, had its roots in 1877 at Queen Victoria's Golden Jubilee when, to celebrate the splendour of the British Empire, the Queen dispatched a regiment of Sikh soldiers from the Punjab region in India to London, capital of her vast reign. Standing ramrod straight and dressed in ceremonial uniforms with all manner of decorative sashes, flaps, epaulets, and medals, these six-and-a-half-foot-tall troopers were an impressive sight, particularly with their gem-encrusted turbans, which added nearly a foot to their height. The Queen was so impressed that she decided to show them off to the commoners – and also to what passed as society – in her other colonies. So, instead of returning to India following the Jubilee, the Sikh regiment left for an extensive tour of Canada.

Though Punjab was but a small part of India, it provided much of the nation's food. Largely hailing from a tradition of farmers, the Sikh soldiers were enchanted by Canada's vast, fertile lands, and many of them, after they mustered out of the military, decided to make their way back and spend what remained of their working lives there.

Their journey, which for a variety of logistical reasons didn't allow direct access to Canada, took them across the Pacific to western Mexico and then north through the United States. Though not legally permitted entry into the United States, thousands of ambitious Sikhs found the border between Mexico and California porous enough to pass through. Many also found work as seasonal farmhands as they gradually made their way north. Some never made it there at all. Not because of hardship, however, but because they were so taken by the beauty and the bounty of the Central, San Joaquin, and Imperial valleys that they remained in California, with large contingents settling in the Yuba City and Fresno areas.

The Sikh pioneers were not allowed to own farms. Still, their farming skills, their work ethic, and their high regard for education and learning made them prized farmworkers by the local landowners. Within two generations, the naturalized sons and daughters and then the grandchildren of these original Sikhs in America either began farming their own land or became partners with those who were already landowners. Today a sizable number of multimillionaire Sikh farmers work the land throughout central California.

This was already the situation in the 1960s, when I first arrived in the San Francisco Bay Area. True to Sikh principles, these farmers and farmworkers were also very charitable and generous to their communities and, over time, were welcomed into the local culture. Theirs was such a success story, in fact, that it was told throughout India where many Sikhs, undergoing year upon year of political strife, came to question whether the social climate might not be friendlier in California than in their own Indian hometowns.

So compelling was this saga of Sikh refugees living and thriving only a few hours by car from where Satinder and I lived in the Bay Area, that one day we drove down to Yuba City just to look around, to get a feel for the place and to connect person-to-person with some of these farming Sikhs we'd only read or heard rumours about.

That turned out to be easy enough to do. Imagine an unfamiliar, citified stranger in a Sikh turban with his attractive wife in her sari sitting by themselves in a coffee shop, or window-shopping down Yuba City's main street, ostensibly waiting for some friendly Sikh to welcome us. And if none did, we'd introduce ourselves to any affable-looking man in a turban accompanied by a woman in a sari. And so it was that, over time, we cultivated a number of friendships.

As to the land itself, it was instantly apparent to Satinder and me what had stopped those turn-of-the-nineteenth-century Sikhs in their tracks during their supposed migration to Canada. It was the central California landscape, which was lush and lovely, reminiscent of Punjab, yet with even wider expanses and broader vistas. Driving through the San Joaquin Valley our first time, I recalled the day my grandmother first taught me about property.

It's not that I, myself, had worked the land as a young man, or that I came from a long line of farmers. I hadn't and I didn't. But somehow, simply by being Punjab-born, I felt a kinship to the soil, the crops, the earthy and fecund smells, and the geometric precision of the groves that grew to the horizon and beyond.

Among our newfound farming friends was a local civic leader, a rich Sikh named Paritem Singh Poonian, and his wife, who invited us down for a few days' stay. One evening, the four of us sat under an arbour behind their sprawling, though unprepossessing, farmhouse. We had just finished our meal; much, if not all, of what we'd eaten had been either raised or grown nearby. It was then that I asked the question that I'd been wondering about since our earlier visits.

'Why, if there are so many Sikh farmers in the region, isn't there a single Sikh temple?'

'Ah, yes,' our host lamented. 'For that, for worship, we have to travel nearly a hundred miles to Stockton.'

'Yes, but why not right here? In Yuba City? Surely there are enough Sikhs in the surrounding towns to support it. After all,' I went on, 'Sikhs have been here for nearly seventy-five years. And still no temple. Why?' Satinder squeezed my hand to suggest that perhaps I was expressing my concern with too much ardour.

'You ask "why?"' our host began, 'and it pains me to admit, "for *not* a very good reason."'

'So, tell us,' Satinder said, folding her napkin and placing it before her on the embroidered table cloth. 'Tell us the reason.'

Our host folded his napkin likewise, then went on to explain. 'You know how argumentative Indians can be, right? Well, picture a gathering of them, older men mostly, sitting, bickering, bickering, and bickering. Not able to agree on *anything* and disagreeing on just about *everything,* from arboreal rights to local politics to employment restrictions. There's never even the time to talk about a temple. Never time to talk about a unifying rather than dividing force. It's simply exasperating!' Our host slapped the arms of his chair, rose suddenly as if to further express the depth of his exasperation, and then, without saying a word, sat back down.

'You know, Narinder,' he began, regaining his composure, 'you know, you could really help us here. A man of science like yourself. A proponent of logic. And, probably most significant, someone from out of town. A 'disinterested' party, so to speak. You could mediate. Get them to agree on… on… on….' He searched for what that 'on' might be, couldn't come up with anything, then finally offered '… agree on at least *something.* And,' he added, 'maybe arrive at some productive approach to start our thinking about building a temple.'

'But I'm *not* disinterested,' I insisted. 'On the contrary, I *am* interested,' I said. 'Very.'

Quiet reigned around the table as our host weighed my remark. 'Yes,' he finally offered. '"Interested." But not so interested that it would cloud your vision or judgment.' He paused, 'Am I making sense?'

'Not quite. But I'm sure we'll figure something out.' My mind raced for what that something might be. But sitting there at the table in that calm Central Californian evening, I couldn't come up with a thing beyond what I already knew: that science was a lot less complex than human relationships.

The arrangement we devised was as follows: Our dinner host would convene two dozen of the most influential local Sikhs, half on one side and half on the other of the primary argument that separated them – an argument, I quickly discovered, that broke down along traditional political lines, liberal versus conservative. Our host's job was to ensure that the sides were well and equally represented and would be seated across from each other in a large room.

Meanwhile, I, the not-excessively-interested man of science and logic, the out-of-towner, would mediate and direct the conversation. It would be my job to, if not actually get people to agree – where there was little historic reason to think they would – then at least to discuss the issues they would agree to disagree upon and somehow forge a peace. All this, while touching on the subject of a temple, as the situation allowed.

A group was gathered. When the exercise began, some of the men – and they *were* all men – were shy and near-silent. Others were voluble and forthcoming. But by the time decades-long differences were broached and aired, everyone was talking. In fact, all that prevented the meeting from continuing on indefinitely was that, going into the process, we had agreed that the talking would stop in precisely two hours, after which I would provide a summation and some insight into how to move forward.

The only problem was that, sitting in that room and listening to these two sides at virtual loggerheads for what was rapidly approaching 120 minutes, I had no idea what I would say, or how I could provide any help or insight whatsoever. And then, just as time was running out, it came to me:

'Listen!' I said, to the low, angry murmur the meeting had devolved into. 'Listen!' I repeated, I insisted. 'After all,' I said, 'I've been listening to you all for two hours now. I've heard you express some kindness, but mostly unkindness. Some generosity, but mostly a lack of generosity. And now it's time for *me* to tell *you* something! And it's this: You might not know it. You might not believe it. But I know it! I believe it! I know that you *want* a Sikh temple here. I know that you *need* a Sikh temple here in Yuba City…'

I paused, that instant knowing that with these two-dozen dubious, stubborn, and argumentative men, I would have to sweeten the pot. I'd have to make them an offer they couldn't refuse. So I did.

'In six months,' I continued, 'Sikhs everywhere are going to be celebrating the 500th anniversary of the birth of Guru Nanak, the founder of Sikhism. If by then you've secured the land for a temple and laid the cornerstone, I will donate $5,000 toward its construction' – a sizable sum back then in the 1960s.

I don't really know why my presence that day was as successful as it was. Maybe it was nothing more than putting my own money on the table. And maybe all the dialogue that proceeded it was just that much chin-wagging. But for some reason, after that day the two camps managed to work together. And indeed, in less than six months there was a ground breaking ceremony in Yuba City for the area's first Sikh temple.

Satinder and I were invited, of course, along with our two children, and I was asked to break ground with the ceremonial first shovel. Since then, millions of dollars have been lavished on that temple, and I am nothing but appreciative of the opportunity to help set the plan into motion.

I found the entire process of working with Sikh farmers to see the temple to completion so gratifying, in fact, that I was inspired to invest in a farm myself. Again, not to work the soil, but to be a part of the growing process, the place, and the culture. It's a powerful idea, I think – farming, *agriculture*, being a part of an effort that grows the food that feeds a nation.

Since then, and over the years, I have been an owner or part owner of a number of farms, growing walnuts, pears, prunes, and grapes. These partnerships and ownerships have led to many delightful day trips to the Imperial, San Joaquin, and Central valleys, often with Satinder or one or more of our kids, for a hearty farmland lunch and an enjoyable drive through the fruit groves. How far I had come from the day I was a boy pilfering oranges with my cousin Jagdish from the maharaja's orchards! And yet, how close I still felt to that time.

Wine and Oil

My most substantial – and, by most measures, my most profitable – venture into farming was my purchase of 3,000 acres of old vine grapes in the Central Valley near Fresno. It was a Friday, I think – late in the week, in any event – that I first heard of a sizable prime vineyard going up for auction that coming Monday. As it stood, California farmland was a solid investment more often than not. And, given the enormous popularity of California wines and the recent upswing in wine consumption across the country, I knew that a mature vineyard would be all the more valuable.

As soon as I heard about the acreage, I asked each of my three current farming partners – Brad Jeffries, Didar Bains, and Milan Mendrick – if they were game to bid on a purchase. Their answer was a resounding and unanimous 'yes!' provided we could first get a look at the acreage to see if it was as advertised. What precisely that meant was unclear to me and the others. Nevertheless, within a few hours we rendezvoused at the Palo Alto airport, chartered a plane and a pilot, and flew to the Fresno area where we did a quick flyover of the coordinates that defined the plot to be auctioned. Satisfied with what we saw, we returned to the Bay Area, ready to commit millions. The fact was, none of us knew the difference between a thriving vineyard and a failing one, particularly one seen from a few hundred feet in the air. But, as it turned out, that wasn't really even the main problem with our reconnaissance flight. Despite having been given the coordinates for the acreage, we ended up flying over the wrong plot. Not the one we bought, in any event.

As for the sales process itself, it was also somewhat out of the ordinary, requiring a 10 per cent earnest-money cash deposit to secure our bid, and a cashier's check for the balance immediately following the auction, which was to be held in San Francisco and presided over by a judge, rather than an experienced auctioneer.

For some reason (I'm not sure why, as I was, by far, the least high-rolling member of our foursome), it was decided that I would be the one to come up with the cash. Fortunately for the deal, I had a solid banking relationship with my branch manager, who, as good fortune would have it, was at work that early Monday morning. About to enter into the largest deal of my agrarian life, I felt quite excited, like-a-kid excited, like I wished I could call and tell Jagdish, my fruit-pilfering partner.

Entering the bank, empty attaché case in hand, I spotted the manager from across the lobby and gave him a quick, jaunty wave. He seemed slow to return it. 'What can I do for you today, Narinder?' he asked as he approached, his hand out to shake mine. But for some reason, he wasn't picking up on my excitement. It didn't seem like him, so I asked him if something was wrong. I wasn't really expecting an answer, and certainly not the one he gave.

'A terrible thing, Narinder,' he blurted. 'A terrible thing,' he repeated, slumping down on his chair. He seemed a broken man. That moment I banished all thought of the money from my mind. With trepidation, I asked him what the terrible thing was.

'My son... some friends... the bridge gave way... he died... just like that.' I felt instantly heartsick, unable to imagine anything worse than losing a child. I walked over to where he sat and put my arm around his shoulder.

'So what can I do for you?' the manager suddenly offered, recovering, a seemingly different man from the one just a few seconds ago. It was as if he had just discovered some deep well of strength within himself.

'No,' I insisted. 'It's too frivolous, what I have to ask.' But he insisted, so, embarrassed, I told him how much money I needed and what I needed it for. Quietly, he walked with me to an officer's desk and asked her to draw me a certificate of deposit.

Back in the lobby, I offered my heartfelt apologies one more time, then thanked him for everything too profusely, I'm sure, deriving precious little joy in walking down the street, even with a cool half-million CD in my case.

On Monday afternoon, all four of us showed up for the auction in San Francisco, thinking that winning the bid would be a slam dunk, given how large the plot was and its high value. Back then, not many individuals could come up with that amount of money in that short a time, even any of my wealthy Sikh friends. A quick look around the room confirmed what I'd anticipated. Aside from the judge and a court officer of some sort, there was only us and another grouping of four individuals across the room – owners, we wouldn't discover until after the auction was complete, of some of the neighbouring acreage.

We opened the bidding with a $1 million bid right out of the chute. This caused a small perturbation in the group across the room, which, after a short powwow, raised the bid by $10,000. We then did the same... as did they... and then us... and then the other group again, back and forth, back and forth, never larger than $10,000 increments.

The bidding went on like this for quite some time, $10,000 at a time, almost to the point of tedium. And we had yet to reach the $2 million mark. Still, the others persisted, mimicking our every move. I was reminded of a recent movie that Satinder and I had seen, *Butch Cassidy and the Sundance Kid*, about two colourful and charming desperados being tracked by a dogged and relentless posse, always on their trail, never far behind. 'Who *are* those guys?' Butch kept asking Sundance as he saw them on the horizon. Except that the question in our case today was 'Who *are* those bidders?'

We were surprised, too, that the judge wasn't encouraging larger increments. But he wasn't. After about a half hour of bidding, I decided it was time to accelerate the process, otherwise we'd either fall asleep, or, after a few more tedious hours, reach the $5 million maximum bid we were prepared to go to. So, I posed a question to my partners: 'Do we want this land or not?' We all agreed that we did. 'Then it's time we make a pre-emptive bid!' I insisted. We all agreed to that, as well. Given the increasing reluctance with each bid our competition made, the very next time the bid came back to us, I countered with a much more aggressive bid. And just like that, the opposition died – and we were the owners of one of the larger grape-growing tracts in the Central Valley.

To celebrate, we decided to take a proper trip to tour our newly acquired property, this time first by car and then on foot, rather than a flyover. It was only after we arrived at the actual acreage that we realized the mistake we'd made a few days ago. The plot we'd identified from the air was, indeed, fully grown in with mature grape vines, as advertised. But *this* one, the *actual* one we paid for, had a number of working oil rigs interspersed throughout the scraggly acreage. In fact, there were ten oil-producing wells on the property – unbeknownst to us, but probably not by those bidding against us.

Not only had we bought 3,000 acres of grapes, it turned out we'd also bought the mineral rights to a producing oilfield! Moreover, as we'd discover when we studied the deed, as part of the sale we'd also purchased the mineral rights to an additional 7,000 acres, for a grand total of 10,000 acres. Farming had suddenly become far more profitable that we'd ever dreamed.

Within two years, we earned back our total investment. Four years after that, we subdivided the acreage into four parcels and sold them at an incredible profit, even as we retained the mineral rights. We still receive monthly royalty checks.

Over the years, meanwhile, the value of my other farms has increased, if not accordingly, then significantly. One 150-acre plot, for example, has risen in value more than eight times over.

Which all leads me to conclude that those fellow Sikhs who settled in the Valley more than a century ago sure knew what they were doing.

Narinder with partners, breaking ground for the first Sikh temple in Yuba City, in California's Central Valley farm region, 1969.

SITE OF NEW TEMPLE — The Sikh community of the Sutter-Yuba area chose the 500th anniversary of the birth of their founder, Guru Nanak, on Saturday to dedicate the site of a new Sikh temple at the corner of Tierra Buena and Young Road. From left, front row, are Kartar Singh Khera, a director of the temple; Ranjit Singh Grewal, leading the prayer; Amar Singh Samra, another director; [illegible] and Puna Singh, another director. Between Mr. and Mrs. Samra is Mrs. Puna Singh and in the back row, right, with the turban is Narider Singh Kapany, another director. The ceremony followed a celebration at Yuba-Sutter Fairgrounds honoring Guru Nanak. Although the only Sikh temple presently in California is in Stockton, the largest concentration of Sikhs in the state is in the Sutter-Yuba area.

Site of New Temple: The Sikh community of the Sutter-Yuba area chose the 500th anniversary of the birth of Guru Nanak to dedicate the site of a new Sikh temple at the corner of Tierra Buena and Young Road. *From left, front row*, are Kartar Singh Khera, a director of the temple; Ranjit Singh Grewal, leading the prayer; Amar Singh Samra, another director. Between Mr. and Mrs. Samra is Mrs. Puna Singh and in the *back row, right*, with the turban is Narider Singh Kapany, another director. The ceremony followed a celebration at Sutter-Yuba Fairgrounds honouring Guru Nanak. Although the only Sikh temple presently in California is in Stockton, the largest concentration of Sikhs in the state is in the Sutter-Yuba area.

PART VII

BEING SIKH

Sikhism: Equality and Charity

You would think that at ninety years of age, I wouldn't change the way I perceived my place in the universe because of a single email. And yet, that's precisely what happened not that long ago. The email in question was from Professor Gurinder Singh Mann, who at the time was holder of the Sikh Studies chair at the University of California, Santa Barbara. Attached was a highly detailed family tree that traced my ancestry directly to the third Sikh guru, Amar Das, a deeply respected man and a major prophet of the Sikh religion.

It's not that, in my long and sometimes contemplative life, I hadn't at least considered that I might somehow be related to one of the ten Sikh gurus that lived between 1469 and 1708. It's just that I'd never paid much attention to my possible place in the larger Sikh family tree. As a moderately religious man and a faithful, practicing Sikh, I was, nevertheless, more a lover of Sikh tradition than I was of its liturgy and religious trappings. I loved, too, the magnificent poetry and art that the Sikh religion spawned – so eloquently celebrated in the two-volume story collection Janam Sakhis that I'd inherited from my father, that chronicles the life of Guru Nanak, the founder of the Sikh religion.

Still, seeing my family tree for the first time, with all the names and relationships meticulously researched and in their proper places, awoke a powerful force in me, connecting me to the past as never before. I set the document down on my dining room table and went to the nearest mirror to see myself, if possibly, anew. But, no. Still the same old familiar fellow. But regardless of what my eyes revealed – or didn't, as was the case – I felt it. I was both today's familiar Narinder and yesterday's distant heir of a prophet, lending me, I dared hope, a whiff of greater wisdom.

Not surprisingly, what I valued first about the Sikh religion and tradition was influenced by how I'd first experienced it. One of my earliest memories was of a meal at the Golden Temple in Amritsar. Hundreds were in attendance, along with my own family, all sitting on the floor and dining together.

'Langar,' Father explained. 'It's the Sikh word for a meal at any Sikh house of worship. And,' he went on, 'it's always free to anyone who chooses to partake. At any Sikh temple in the world. For anyone who comes for whatever reason. No one needs to leave hungry, in body or in spirit.'

Because Father was its source, I accepted this truth for much of a year, assuming the motivation behind it as *charity* and concluding that we Sikhs were good and giving people. And practical, too. After all, it's harder to pray on an empty stomach than a full one.

But when I shared my insight with Father one day at our local temple, he said, 'Yes, by all means, we Sikhs are charitable people – and that's certainly a *part* of our tradition – but that's not the only reason that Guru Nanak instituted the practice of providing a free meal to everyone in attendance. No, that wasn't only for charity's sake,' Father insisted, becoming more and more animated. 'Rather, it was to demonstrate another primary tenet of Sikhism: *equality*. Absolute equality, among men and women, the young and the old, the rich and the poor.'

Now, for me, still a young boy, charity was a relatively easy concept to grasp, particularly as I had seen it practiced by my generous and charitable family in our community. The notion of equality, however, was a bit more perplexing. Equality between the sexes? Possibly. I recalled passages in the Guru Granth Sahib, our holy book, that celebrated it. Equality between the ages? Maybe, yes. But equality between the rich and the poor? How was that even conceivable? I harkened back to the visits with my mother and aunts to the peasants in the villages neighbouring our ancestral home. I confessed my dubiousness to my father.

Father smiled. 'It's true, Bino – equality even between the rich and the poor.' He went on to tell me the story of a Mughal Emperor who arrived one day to pay homage to Guru Amar Das, one of our Gurus. And when it came time to eat, the emperor was asked to sit on the floor with everyone else, was served the same food, and ate in the same room. And this was not in the 'modern days,' Father went on, 'it was hundreds of years ago when our religion was still very young.'

Some fifty years after this talk with Father, Satinder and I, along with our children, Raj and Kiki, witnessed this unique aspect of Sikhism in another memorable experience. On this particular day, a special Sikh day of celebration, the four of us were travelling through the Punjab countryside where Sikhs in food stalls lined the road in virtually every village, offering free food and drink, no matter how seemingly meagre their own circumstance.

'Which is not to say that charity isn't important to Sikhs. Because it is! Extremely so!' Father insisted way back then. 'Extremely so!' Bearing out Father's insistence, now some eighty years since we had first discussed it, is a recent British study that revealed that Sikhs are proportionately the most charitable group in Britain, followed by Jews, and then by all the others. The results were reassuring, somehow, even as they stopped me in my tracks, prompting another long-ago memory when, once again, Father was the central character...

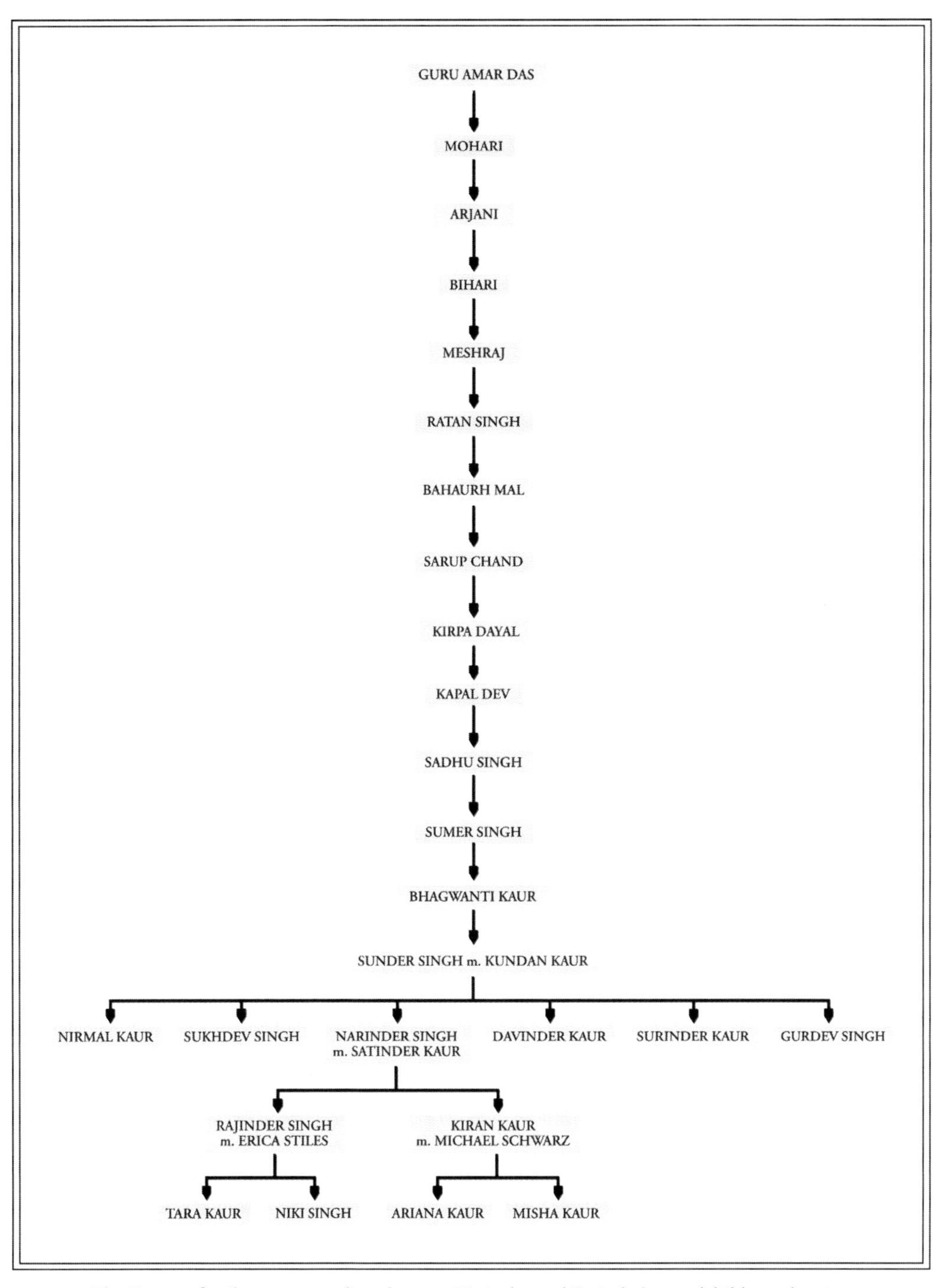

The Kapany family tree, extending down to Narinder and Satinder's grandchildren, showing their descent from Guru Amar Das, the Third Sikh Guru (1479–1574), Punjab.

Guru Amar Das, Northern India, *ca.* 1800–1810, opaque watercolour and gold on paper, 17.6 × 25.8 cm, Kapany Collection.

The Art of Giving

It has just begun to rain. A soft rain in an early spring evening. I am about twelve or thirteen – it is the late 1930s. Sitting outside in deck chairs on the veranda under a faded yellow awning is my father and a favourite walking companion of his, Mr. S. Mota Singh. I have been sick that day, the flu or something, and the illness has made me particularly attuned to the sounds of the encroaching night, my father's voice, especially. The two men are discussing what they might contribute to the community.

'Perhaps we can build a Sikh meeting hall,' Mr. Singh suggests. But Father says, 'No. Too many people will go there to argue and get raucous.'

Father then suggests a hospital, but no, both men quickly decide: A medical facility will be too expensive to maintain. Finally, after considerable back-and-forth, the two men agree to build a small elementary school to serve the children of the community.

As it turned out, that small building was merely their initial contribution. Over the years – and largely through the benevolence of these two men – that simple elementary school provided the foundation for a larger middle school, then a high school, and finally a small college, where Father, staying on as an emeritus dean well into his nineties, signed documents and checks until he passed on. All in the spirit of contributing something of value to the community.

That was, after all, the charitable impulse that led to the construction of that first small school. It was an impulse that Father nurtured his whole life, and that I am grateful to have inherited – if, indeed, such a trait can be passed on genetically – from the man I remember on that rainy spring evening.

As for myself, I have worked hard at upholding this tradition, trying to be a generous and charitable man, most often in ways that reflect my beliefs in the value of education and in the insight and enrichment that comes through art. I have endowed a chair of Sikh Studies in my mother's name in the Department of Religion at the University of California, Santa Barbara. I've also endowed two chairs at the University of California, Santa Cruz, one in optoelectronics, the other in entrepreneurship.

Recently, too, I'd undertaken the process of packing up and donating almost my entire English library of Sikh books – many of them art books, over 1,000 volumes in all – to UC Santa Cruz in my father's name, where a sizable and elegantly appointed reading room has been inaugurated and is seeing regular use.

Father always said that when you're young, you're full of piss and vinegar, working hard to get there, wherever 'there' was. You have some successes and some failures and then you get to a point where you say, 'Okay, I have enough now to comfortably take care of my wife, myself, and my family. Let's see what I can do for others.'

Beyond books on Sikh art and heritage, about sixteen years ago I also donated a hundred pieces of Sikh art – including paintings, textiles, and other art pieces – to the Asian Art Museum in San Francisco, and have underwritten the creation of a permanent gallery of Sikh art in the museum in my wife's name. I also have given the museum one of the two illustrated Janam Sakhi manuscripts that my family has treasured for generations.

I'm also currently in negotiations with a number of other institutions to determine the best home for another major donation of Sikh art from my extensive collection. And, even after the decision is made and the paintings and sculptures are sent, I'll still have dozens of priceless pieces on the walls of my home and my office to feed my own hunger for fine art, most of it celebrating the Sikh religion and its traditions over the generations.

Left: Narinder speaking at the inauguration of the Narinder Kapany Professorship in Entrepreneurship at University of California, Santa Cruz, with Chancellor George Blumenthall looking on. *Right:* Dr. Rahuldeep Singh Gill, Dr. Gurinder Singh Mann, Dr. Mark Juergensmeyer, Dr. Kapany, and Dr. Pashaura Singh, at a Sikh Foundation conference, Stanford University, 2015.

Left: The 'Satinder Kaur Kapany Gallery' (named in honour of Narinder's wife), Asian Art Museum in San Francisco, inaugurated in 2003; the museum houses one of the world's most comprehensive Asian art collections.

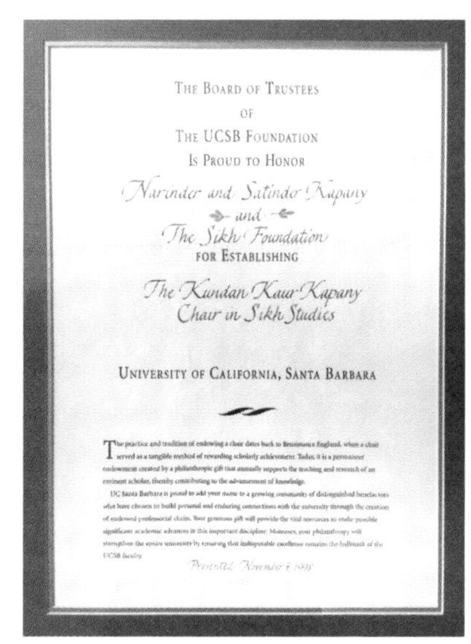

THE BOARD OF TRUSTEES
OF
THE UCSB FOUNDATION
IS PROUD TO HONOR
Narinder and Satinder Kapany
and
The Sikh Foundation
FOR ESTABLISHING
The Kundan Kaur Kapany
Chair in Sikh Studies

UNIVERSITY OF CALIFORNIA, SANTA BARBARA

The practice and tradition of endowing a chair dates back to Renaissance England, when a chair served as a tangible method of rewarding scholarly achievement. Today, it is a permanent endowment created by a philanthropic gift that annually supports the teaching and research of an eminent scholar, thereby contributing to the advancement of knowledge.

UC Santa Barbara is proud to add your name to a growing community of distinguished benefactors who have chosen to build personal and enduring connections with the university through the creation of endowed professorial chairs. Your generous gift will provide the vital resources to make possible significant academic advances in this important discipline. Moreover, your philanthropy will strengthen the entire university by ensuring that indisputable excellence remains the hallmark of the UCSB faculty.

Presented November 7, [illegible]

Above: Narinder at the 2017 inauguration of the 'Sundar Singh Kapany Book Collection' and the 'Sundar Singh Kapany Group Study Room' (named in honour of his father), McHenry Library, University of California, Santa Cruz. *Left:* In commemoration of the 'Kundan Kaur Kapany Chair in Sikh Studies' (honouring Narinder's mother), University of California, Santa Barbara, 1998.

Sikhism: A Spiritual God and an Equal Opportunity

While equality and charity are the two primarily outward-facing, most often physically manifest tenets of Sikhism, the religion's primarily inward-facing tenet is a belief in a *formless* representation of God. Which is to say, the Sikh concept of God is exclusively a spiritual one, one that never takes on a human or a mythological shape. Ours is a God who never walks the earth, who never asks to be seen. Consequently, as a man of science, I don't have to make impossible demands on my imagination to realize Him, nor does it beg my credulity to allow Him into my life.

Another tenet of Sikhism I find consonant with and reflective of my own personal values as a human is that everyone should be given the opportunity to rise from whatever level he or she is on, to whatever level his or her skills, capacities, and spirit enable. In a subcontinent like India that is primarily caste-driven, Sikhism doesn't recognize castes.

Which is not to say that every Sikh practises every or, indeed, any of these basic beliefs. There are proportionately as many bad actors among Sikhs as there are in any other religion. But the good ones often lead charitable and hard-working lives.

Sadly, the non-Sikh world isn't altogether tolerant of Sikhs, no matter how good their works. Having lived in liberal academic and entrepreneurial environments much of my adult life, on three continents, I've also managed to avoid the prejudice that many Sikh men have had to endure, often for their perceived 'otherness' as manifest by their turbans and long beards, which also often lead to the misguided assumption that we are from the Middle East. It's not that there is anything wrong with being Middle Eastern, it's only that Sikhs come from another place entirely, with a different culture, religion, and worldview. We are our own people, yet we endeavour to blend in to the cities and communities where we live, anywhere in the world.

Top, left: Guru Gobind Singh, Northern India, *ca.* 1830, opaque watercolour on paper, 18.5 × 15.2 cm, Asian Art Museum of San Francisco, gift of the Kapany Collection. *Top, right: Guru Nanak*, Sobha Singh, 1969, oil on canvas, 56 × 71 cm, gift of Dr. Janmeja Singh to the Kapany Collection. *Bottom, right*: *Portable Palki* (palanquin), with a miniature Guru Granth Sahib, Northern India, silver with ink on paper, 13.3 × 6.4 cm (palki), 2.3 × 2.5 × 2.5 cm (holy book), Kapany Collection. *Bottom, left: Camp of Bhai Veer Singh*, Punjab, *ca.* 1850, opaque watercolour on paper, 54.9 × 37 cm, Kapany Collection.

PART VIII

COLLECTING AND SHARING SIKH ARTS

The Gift of Art

Of the gifts I've made to others and the gifts I've made to myself over the years, the largest, by far, has been the gift of art.

Why art? Though I'm not really sure, I think it goes back to when I was a boy in the 1930s, when India had precious few museums, rare venues where the young, in particular, could experience art. As a result, I was brought up mostly without it, except for the paintings on the walls of my home and the ancestral homes where my grandparents lived, and in the pages of those two magnificent, illustrated holy books that my grandmother generously gave to my father, and then from my father to me.

Whether that explains my enormous, subsequent lifelong interest in art, Sikh art in particular, I don't know. All I know is that I came to it with a substantial hunger and no small amount of appreciation. I especially admired the paintings, the beauty of the images themselves, the richness of their tones and colours, the complexity of the stories they told, and of course the Sikh culture they reflected. So important were these things to me that as an adult I sought to fill my home with them and share them with others, young and old alike.

It was an interest that took form for me following my move to London in the 1950s where I found myself only a few minutes' walk from three leading museums – the Victoria and Albert Museum, the Science Museum, and the Natural History Museum – as well as the antique and historic holdings of Imperial College, which I'd attended.

Not one to do things in a piecemeal manner, I didn't simply fall in love with a single painting or a painter or two; I fell in love with it *all*, the whole, often-rarefied world of museums, the carefully curated collections and exhibitions, the arcane workings of auction houses, the odd and quirky art dealers, the special shops in out-of-the-way neighbourhoods, and even the local street fairs and art festivals.

Having said that, back then – and since – I was *particularly* enchanted by the work of an Indian painter named Amrita Sher-Gil. Born of a Punjabi Sikh father and a Hungarian Jewish mother, she was the single artist most responsible for bringing modern representational painting to India. Most of her work depicted rural Indian life – people living and working. Over only a few short years, interest in her work grew so great that she was dubbed the 'most expensive female artist in India'. I've never owned any of her work, though not for lack of wanting. But her output was limited to a few hundred pieces, and she died in 1941, at only age twenty-eight. Fortunately, by the time I had any discretionary money to spend on other, less-celebrated Sikh artists, their work was still quite affordable, and so I started buying all manner of Sikh oil paintings, watercolours, and sculptures – eventually to the point where I was considered by those in the Indian art world as a serious collector.

Throughout much of the twentieth century, little was known about Sikh art, as distinct from the larger category of Indian art issuing from the humongous Indian subcontinent. To provide a deeper look at and a greater appreciation for the works of Sikh artists in particular, in 1974, in the Sikh Foundation's quarterly magazine, *The Sikh Sansar*, I published a special segment literally introducing the term 'Sikh Art.' While the article wasn't widely read, it *did* help capture the interest of a number of the world's major auction houses, notably Sotheby's and Christie's. It's not that these two venerable institutions were led to the discovery of Sikh art through my publication, but having found me to be a genuinely interested investor and full-time champion, they now knew that any Sikh art they might uncover would likely have an interested buyer, or at least someone to give it more than cursory attention.

This virtuous circle resulted in a number of books being written about Sikh art, including one that editors Paul Michael Taylor, a curator at the Smithsonian Institution, and Sonia Dhami, executive director of the Sikh Foundation, published in 2017 under the imprint of the Foundation, titled *Sikh Art from the Kapany Collection.* Featuring commentary by fifteen of the world's most renowned Sikh art scholars, this elegant, profusely illustrated, 320-page volume examines a wide range of Sikh artistic expression and Sikh history and cultural life, as illuminated in Satinder's and my collection, much of which has appeared in exhibitions around the globe.

This growing interest in all things Sikh notwithstanding, I believe that far too little attention is still being paid to Sikh art in countries around the world. Take, for example, the museum in Pakistan that Satinder and I and some friends visited about ten years ago. To its credit, the museum was exhibiting dozens of European paintings depicting Sikhs and Sikh life – both large and smaller pieces, rich in detail, perhaps 100 in all. It was a substantial collection. However, the museum had no apparent temperature or humidity control, and seeing the poorly preserved, shoddily cared-for paintings brought tears to my eyes. I subsequently suggested to both the Pakistani prime minister and the provincial governor to move the paintings to a larger, more central museum, but it didn't happen. As far as I know, the paintings continue to deteriorate.

Meanwhile, spanning about a half-century of collecting, I've purchased at least 500 pieces of Sikh art at auction and a few hundred more from galleries and private collectors. And this deep interest in art and my impulse to collect it began with those two richly illustrated volumes that my grandmother used to read from to her family.

The Koh-i-noor

Fresh ideas come to me all the time: One morning, about fifteen years ago, for example, while brushing my teeth at the same time that I was performing my twice-daily stretching exercises, I recalled that I had once been deeply moved by a painting of Rani Jind Kaur, the wife of the great Sikh Maharaja Ranjit Singh. So deeply moved had I been that, that very morning, I determined to view the lovely Rani's countenance daily. The problem was that the painting of the Rani of which I was so enamoured resided in the private collection of an American, Dr. William Ehrenfeld, a significant collector of Indian art who likely appreciated the painting as much as I and who would be loath to part with it.

That wasn't the case, however. When later the man and I spoke over the phone, he confided that just the day before he had shipped the painting to Christie's in New York to sell at auction. Just imagining the painting in transit – and *not* to me – caused my tender heart to skip a beat.

'But,' I said to the good doctor, 'the economy is in bad shape. People don't have enough discretionary money for such a purchase these days. Better to hold on to it for a year or so. *Or*' – and I said it as if the idea had just come to me – 'better yet, I could take it off your hands this very day. And,' I added, 'for a good price.'

The doctor chuckled at my feeble attempt at manipulation. 'What "good price" would that be, Dr. Kapany?'

I 'doctored' him back in my response, according him his full title, as he did me. I gave him my offer. There was no laughter on the other end of the line now.

'That's an eminently fair offer,' he acknowledged. 'Quite so,' he said, obviously considering it.

'Yes,' I agreed, the clear hint of a transaction in the air. 'It's a beautiful painting. I will give it a good home.' He told me that he needed an hour to talk to Christie's, then rang off.

Promptly an hour later, we were back on the phone. The doctor had called Christie's and, pending an inked deal between the two of us, they agreed to ship the painting directly to me rather than ship it back to him. 'I've already parted with her, said my goodbyes,' the doctor said by way of explanation. 'I don't think I could do it a second time.' And he didn't.

Today, after being loaned for display to the Asian Art Museum in San Francisco and the Smithsonian Institution in Washington, DC, that painting of Rani Jind Kaur hangs in my living room. It would probably fetch upwards of millions at auction, though it is worth far more than that to me.

Duleep Singh was Rani Jind Kaur and Maharaja Ranjit Singh's son, nine years old when he was taken from the Punjab by the British. He would also become the adopted son of Queen Victoria, who, when the British annexed the Sikh Empire in North India, ordered the boy to be brought to her in London. Also part of the plunder that found its way to London and Queen Victoria, was the renowned Koh-i-noor diamond, the largest diamond in the world, with a price above priceless.

The story has it that rather than hiding the gem away to keep it safe, Maharaja Ranjit Singh wore it regularly. But not to adorn some gem-encrusted ceremonial robe or uniform, or even in concert with other lesser baubles and medals. Rather, he wore it as the sole adornment on the simplest of white suits or robes while others around him were often dressed to the hilt.

To make the Koh-i-noor truly hers, some years after she received it, the Queen sent the diamond to Brussels to be recut. This took quite some time, as recutting a gem of such beauty and great value required all the diamond cutter's skill. After some time, it was finally returned to her.

Duleep, meanwhile, had become a significant presence in Queen Victoria's court, and grew up under the tutelage of a lady of the court, Lena Campbell Login.

Again, as the story continues to have it, on the day the diamond was returned to the Queen, Duleep, then about sixteen years old, was standing on a platform posing for a painting by a European artist, Franz Xavier Winterhalter, whose masterwork-in-the-making would eventually become a priceless piece of Sikh art.

When the Queen, holding the diamond out in her palm, asked the young man if he wanted to see what was once his father's prized diamond, Duleep stepped off the platform and, without a word, took it from her, walking with it to the window as if to see it more clearly in the natural light of day. It was summertime and all the windows of the palace were open. For a few seconds, Duleep stood with the diamond in his hand, holding it more outside the window than in, gazing out beyond the palace grounds and to the horizon. It was not written in any of the stories I've read about this event, but I imagine that in those moments Duleep remembered the diamond attached with a silk armband and tied around his father's upper arm – that man, that great maharaja, who at one point had conquered all of North India.

Whether Duleep actually saw this in his mind's eye or not, the nervousness must have been palpable in the room and among the members of the Queen's court. Everyone was surely anxious that the maharaja's son might simply, spitefully, fling the newly cut diamond out the window and it would never be found. He didn't, however, and instead handed it gently back to the Queen and returned to his pose on the platform, completely nonplussed, with no indication that moments before he'd held the world's most precious gem in his hands – a gem that, to this day, remains in the possession of the Queen of England.

While the painting of Duleep that was being created that day was never available to me, not long ago a small sculpture of him was about to go up for auction in London at Bonhams, a well-known auction house. Satinder and I travelled to London to examine it. A Bonhams representative led us to a small room where the statue sat on a small table. It was not without defect, yet I still felt that the £35,000 floor was quite low. In fact, I thought the piece would go for a lot more. Trying not to reveal my eagerness for the sculpture, I offered to meet their floor amount if they would pull it from auction. I even took out my chequebook to show them I was ready to close the deal. The slightly flustered Bonham representative helping us (apparently, accepting such an offer was above her pay grade) said she'd have to check with her superior. I immediately suspected this would not bode well for my offer. I was right. Arriving on the scene, the boss explained that the sculpture had already been advertised and would, therefore, have to go up for auction, no matter how convenient or generous my offer was to take the sculpture off their hands. Suffice it to say, I left Bonham's disappointed. Even more so when, participating in the actual auction over the phone from San Francisco – bidding as high as £100,000 – I listened, incredulous, as the winning bid reached 2 million pounds.

Some years later, two Sikh paintings came up for auction at Christie's: The first was of Maharaja Ranjit Singh and his entourage; the second was of the Golden Temple. Desiring them both, I travelled to London to examine the two pieces in person, then made Christie's an offer that I felt would be sufficiently attractive to win their attention. But, no deal. Instead, I ended up paying £47,000 at auction for the Golden Temple painting. I also bid £200,000 for the piece with Ranjit Singh and his entourage, but was substantially outbid, despite having a friend inside Bonham's who did his best to look out for my interests.

Undaunted, my interest in Sikh art was not yet slaked. Even today, many years later, I continue to scour the auction catalogues looking for a piece of the quality, the complexity, and the sheer beauty of the Rani Jind Kaur portrait, which provides me immense pleasure.

Clockwise, from top left: Sikh Art from the Kapany Collection, published by the Sikh Foundation and the Asian Cultural History Program of the Smithsonian Institution, 2017. *Maharaja Duleep Singh*, reproduction by Sukhpreet Singh of the painting by Franz Winterhalter, Punjab, 2005, oil on canvas, 94 × 182 cm, Kapany Collection; *Maharani Jind Kaur*, George Richmond, 1863, oil on canvas, 58 × 75.5 cm, Kapany Collection; Maharaja Ranjit Singh, Fakir Charan Pareeda, 2014, bronze, 22 × 79 × 56 cm, Kapany Collection.

Clockwise from top: Narinder at home, with a marble bust of Maharaja Ranjit Singh against the backdrop of a painting of Guru Nanak by Arpana Caur; Narinder with artists Devender Singh and Arpana Caur at the latter's studio in Delhi; Narinder with artist Sukhpreet Singh during his visit to the Sikh Foundation in Palo Alto; displayed behind them is his painting of the Golden Temple, 2017; The 'Twins' artists Amrit and Rabindra Kaur Singh, at the opening of the Satinder Kaur Gallery, Asian Art Museum, San Francisco.

The Sikh Foundation: Inspire, Educate, Enrich

Father's most notable contribution to his community? Without a doubt, his school and his college. And mine, to my community? Beyond my scientific contributions, it has most certainly been creating the Sikh Foundation… and nurturing it to serve a worldwide community.

The notion hit me back in 1967. I had just taken Optics Technology public, and I had some money – considerably more than I needed, in any event. I pondered the situation: I was only forty, I'd achieved a significant degree of success in the business and scientific arenas, and I was still enjoying my Da Vincian lifestyle, tasting a little of this and a bit of that from the savoury buffet of life. What next? Travel? Philanthropy? Mentoring?

And then it came to me: I would create and fund a foundation that could serve as an umbrella for all my best Sikh-driven impulses – my Sikh art collector's zeal as well as my desire to show and share that art more broadly; my personal belief in the value of art in education; my Sikh-inspired commitment to charity; my pride in the Sikh religion, particularly its humane and ethical aspects; and my desire, again like that of my father, to make a lasting contribution.

And so it was that the Sikh Foundation came into being. It was founded by myself along with twelve trustees, who, in addition to helping provide the Foundation with a focus and a vision, all contributed to an endowment to make the Foundation's good works possible. Remarkably, its mission back then was almost identical to what it is today: to promote and preserve Sikh art, heritage, education, culture, and religion.

To this end, the Foundation works to pass on the Sikh heritage to the growing Sikh diaspora in the West, particularly to Sikh youth, while introducing the non-Sikh world to our ethics, arts, and literature. It also urges Sikhs and non-Sikhs alike to adopt a Sikh perspective as a means for approaching common human concerns, and it helps advance the Sikh tradition of critical and creative thinking that gave birth to the faith.

The Foundation has also played a role in modern Indian politics, albeit a tangential one. In 1974, concerned that Sikhs were not receiving their fair share in independent India, I rang up Maharaja Yadavindra Singh – then the Indian ambassador to the Netherlands, and the very maharaja who had brokered my meeting at the United Nations with Krishna Menon years before – to see if perhaps, as an influential member of the government, he could help. At stake were water problems, Punjabi language issues, and industrial and other concerns. We agreed to meet at his home, where I first thanked him for all the good work he'd already done on Sikhs' behalf. Still, I told him, I felt that under his leadership there was an opportunity to get an even better deal for Punjab. He agreed; and after a solitary, 10-minute walk around his garden, he returned to suggest that we organize a meeting in Delhi of a number of highly regarded Sikhs to discuss the Sikh situation in Punjab. And, to prevent drawing attention to the political nature of the meeting, that we do it under the cloak of the Sikh Foundation. Unfortunately, he passed away not long after and the meeting never took place. Years later, when his son, Amarinder Singh, visited San Francisco, he stayed at my home, and I even took him sailing in my boat. We finally had the conversation I had never been able to have in Delhi about Sikh issues. Singh is today the chief minister of the state of Punjab.

Turning fifty in 2017, the Foundation continues to contribute significantly to realizing its objectives. It has underwritten, mounted, and/or provided the art for numerous world-class Sikh art retrospectives, museum exhibitions, and lecture series, from London to Toronto and the United States, including New York City, Washington, DC, (at the Smithsonian for five years), Texas, and California where, at San Francisco's Asian Art Museum, I have donated the nation's first permanent Sikh gallery – again, because of our commitment to education and our belief that museums are among the best teaching resources.

The four university chairs that the Foundation has underwritten, all in Sikh or Punjabi studies, or both, are testament to our commitment to teaching, as is its support of Punjabi language programmes at Columbia University, Stanford University, and the University of California, Berkeley.

The Foundation has also published a number of books – in addition to the aforementioned *Sikh Art from the Kapany Collection* – including children's books on subjects that include the Sikh religion, history, art, and culture, as well as two journals, newspapers, greeting cards, and fine art posters and calendars.

The Foundation's website features current Sikh-related news as well as an online publication of the *Sikh Research Journal*. Offline, the Foundation's reach takes us all the way to India where, in Punjab, in collaboration with UNESCO, it has funded the first phase of the renovation of the historic Guru Ki Maseet, and where, also in Punjab, it supports the Satnam Sarab Kalyan Trust in schooling over 150,000 children in Sikhism.

Non-profit and apolitical, the Foundation is all the more mindful of real-world concerns, gearing up for the future with both an economic and a technological focus. We are, for example, exploring the costs and benefits of providing and securing universal digital access to Sikh – and other – art collections around the world, as well as devising ways to preserve and continue to hang in museums many splendid Sikh paintings, despite their relatively fragile shelf life and their relatively high susceptibility to the damaging effects of heat and light.

A responsible steward for the artworks in its possession, and well-attuned to the encroaching demands of tomorrow, the Foundation is vigorously responding to the changes as well as the challenges that will transpire in the years to come.

On 5th May through the 7th, 2017, the Sikh Foundation celebrated its first fifty years of service and gave a proper welcome to its next fifty. Attended by guests from all over the world, including Canada, the United Kingdom, Europe, and India, the celebration kicked off with a glorious gala that included a tour of the San Francisco Asian Art Museum's exhibit titled *Saints and Kings: Art, Culture, and the Legacy of the Sikhs*, a cocktail reception, and a dance performance, all capped off by an exotic Indian dinner service along with a presentation of awards to five women who have made significant contributions to the Sikh world. Instrumental music and a second dance performance closed out the evening's festivities.

In the two days that followed, our guests – numbering over 200 – were also treated to a comprehensive program at Stanford University, titled 'Advancing Sikhs with Education,' which featured as speakers some of the Sikh world's most distinguished authors, historians, professors, curators, physicians, technologists, and entrepreneurs.

In 2017, as in the first half-century of its existence, the Sikh Foundation continued to explore the unique Sikh sensibility while honouring Sikh art and artists – the impressions they left, the difference they made, and the lives they touched.

At the Sikh Foundation's 50th Anniversary Gala, held at the Asian Art Museum in San Francisco, May 2017: *Clockwise, from top left:* Dr. Nikky G.K. Singh; Mr. Randeep Singh Sarai with Valarie Kaur; the Hon. Harjit Singh Sajjan with Dr. Anarkali Kaur; Bibi Inderjit Kaur with Arpana Kaur; Mr. Tarlochan Singh with Susan Stronge; Dr. Kapany addressing the gathering.

About us

Established in 1967, the Sikh Foundation is a non-profit and non-political organization for the promotion and preservation of Sikh art, heritage, education, culture, and religion.
Our objectives are:

- Pass on the Sikh heritage to the growing Sikh diaspora in the West, particularly the youth.
- Introduce the world to the ethics, mysticism, arts, literature and heroism of the Sikhs.
- Contribute Sikh perspective to common human concerns.
- Advance Sikh culture by advancing the tradition of critical and creative thinking that gave birth to the faith.
- Generate high quality resources for the study of Sikhism.

Some of our highlights include milestones in area of Arts & Cultures, Academia, and Education.

Raja of Patiala on Elephant, chromolithograph by Emily Eden, 1844. Kapany Collection.

Arts and Culture

1982 Splendors of Punjab: Art of the Sikhs
Asian Art Museum, San Francisco.

1999 The Arts of the Sikh Kingdoms
Victoria & Albert Museum, London.

1999 The Arts of the Sikh Kingdoms
San Francisco and Toronto.

2003 Satinder Kapany Gallery of Sikh Art
The 1st and only permanent Sikh gallery
Asian Art Museum, San Francisco.

2004 Sikhs: A Legacy of the Punjab
The Smithsonian, Washington D.C.

2005 Victoria & Albert Museum Annual Lecture Series
London (2005 – 2008).

2006 I See No Stranger - Sikh Art & Devotion
The Rubin Museum of Art, New York.

2008 Sikhs: The Legacy of the Punjab
Museum of Natural History, Santa Barbara.

2012 Sikhs: The Legacy of the Punjab
Fresno Art Museum, Fresno.

2015 Sikhs: The Legacy of the Punjab
Institute of Texan Culture, San Antonio.

Online Resources

The Sikh Foundation's website ***sikhfoundation.org*** features current news and an online publication of the Sikh Research Journal. In addition, an online store provides a wide collection of books, art, and other products related to Sikhs & Sikhism.

The Sikh Foundation supports Satnam Sarab Kalyan Trust in Punjab wherein instructions on Sikhism are imparted to over 150,000 school children.

Above and facing page: Sikh Foundation brochure celebrating the 50th anniversary of its founding, 2017.

"The Holy Temple" from Original Sketches of the Punjab by a lady. 1854. Kapany Collection.

Heritage Conservation

The Sikh Foundation funded the first phase of renovation of the Guru Ki Maseet at Sri Hargobindpur in Punjab, India in collaboration with UNESCO.

Sikh Fine Art Calendar

The Sikh Foundation promotes Sikh art and contemporary sikh artists through it's annual Sikh fine art calendar. In it's 15th year of publication, the calendar continues to be a collector's item.

Academia

1999 The Kundan Kaur Kapany Chair for Sikh Studies
University of California, Santa Barbara.
2008 The Dr. Jasbir S. Saini Chair for Sikh Studies
University of California, Riverside.
2011 The Sarbjit S. Aurora Chair for Sikh Studies
University of California, Santa Cruz.
2015 Establishment of the Sikh & Punjabi Language Studies Program,
The Graduate Theological Union, Berkeley.

The Sikh Foundation supported Punjabi Language Studies Program at Columbia University, University of California, Berkeley and Stanford University.

2013 Sikh Foundation Calendar.

Publications

The Sikh Foundation has published a dozen books on subjects ranging from religion, history, art and culture of the Sikhs including children's books.
It has also published two journals, newspapers, greeting cards and Fine Art posters.

Rani Jind Kaur, oil painting by George Richmond, 1863. Kapany Collection.

List of our publications
1968 The Sikhs and Their Religion
1969 Guru Gobind Singh: Reflections and Offerings
1972-1977 "The Sikh Sansar"
1984-1985 "The Sikh Times - Echo Of The Sikh World"
1999 Sikh Art & Literature
1999 Warrior Saints
1999 Bindhu's Weddings
1999 The Boy with Long Hair
1999 Baba Ditta's Turnip
1999 Arts of the Sikh Kingdoms
2001 The Name of My Beloved
2006 I See No Stranger
2012 Games We Play

Seal ring of Maharaja Ranjit Singh. Carved emerald set in gold. 1812-1813. Kapany Collection.

PART IX

RULE, BRITANNIA!

A Home with a View

London. So different from any city I'd ever been to before. I was enchanted, smitten with her from the first day. I loved how urbane she was, how sophisticated, how much she had to teach me when I was a young man, and how much more she had to show me as I grew older. I've since travelled to dozens of cities, many of them capitals, but none quite holds the allure for me that London did… and still does. So in 1967, at only age thirty-nine, I staked my claim there.

Drawn to the city as powerfully as I was, and assuming that there would be a growing contingent of others who would soon discover and also fall in love with her – thanks in no small part to the Beatles and the Rolling Stones and Carnaby Street – I bought twelve separate flats in Chelsea and South Kensington as an investment.

The investment was a sound one, I was certain, but realizing any return was a long time in coming. The primary reason for this was that the management company I was using to oversee these properties was largely inept, if not totally dishonest. It was therefore rare that I would see a penny of rent, but rather merely excuses why there wasn't any.

And so, after holding the properties – and my tongue – for nearly thirty years, in the mid-90s, I decided to cash out. I was simply too busy to ride herd on, and try to build trust with, another management company. I was pleasantly surprised, however, to discover that the average worth of the flats was $250,000 each. Rent or no rent, it was an awfully good return. On the other hand, if I had held them instead of selling them, today they would be worth about $10 million.

As fate would have it, I was in London when the real estate agent I was working with closed the last sale. Signing the final documents turned out to be a sentimental moment and, without thinking much about it one way or the other, I asked him to find a very special flat for me, not as an investment, but for a place to stay anytime I had a yearning to be in London. Again, without giving it much thought, I specified areas of the city I'd favour and the general ambience I was looking for. Neither of us ever mentioned money, with the mutual, tacit understanding, I believe, that if he found the ideal place, I'd pay what it would take to get it.

What followed was a flurry of looking at flats throughout the city, none of which caught my eye or satisfied my inner London fantasy. They were either too small, too musty, too impractical, too dark, or too far from where I'd want to spend my time. After about a week of frantic searching, I grew tired of the hunt and called the agent off. Immediately after we spoke I made a reservation for a flight back to California the following day.

The next morning, while I was waiting for a car service to take me to the airport, the agent called. I took it on the house phone in the lobby of the Intercontinental Hotel. 'I believe I've found the flat of your dreams,' he began. I didn't know whether to take him seriously or not. Sensing my indecision, he insisted, 'Truly!'

Already in the lobby with my bags huddled around me, my mind no longer on flats, I, once again, tried to call him off. But the agent was dogged – I like that in a person in that profession – and told me he'd swing by to pick me up. We'd check out the flat together, and then he'd drop me at the airport.

Fifteen minutes later the agent and I were driving down Exhibition Road, nearing Imperial College and its own three famous museums, my luggage safely stowed in his car's boot. The flat was in a lovely building owned by an Arab gentleman of considerable means. Standing at the door to the flat, the agent doubled down on his assertion that this was the flat of my dreams. I was dubious, but not altogether dismissive.

As it turned out, he was right on the money. The apartment was perfect. Impeccable, exactly what I wanted. And right on Exhibition Road. I suppose I should have considered it more carefully, or, at least, talked it over with Satinder. But there was no time to dally. An apartment as princely as this wouldn't stay on the market long. Besides, my plane was leaving in a little more than an hour and a half and we were a good half-hour away in traffic from Heathrow Airport.

'Well...?' my agent asked.

'Well, I *like* it!' I was both eager about the flat and anxious about making my plane. Having spent no more than 5 minutes looking the place over, I asked, 'How much?'

'Fifty thousand,' he said. 'Pounds.' I appeared to contemplate the amount, though my decision had already been made – a cursory one, to be sure, as I was not even certain where the flat was located in the building, or precisely where the building was on the block, or what the flat's windows overlooked.

'I'll take it!' I said.

'Just like that?' the agent asked.

'Just like that.'

A few months later, I returned to London, forgoing my regular reservation at the Intercontinental in favour of staying in my own new flat. The place, after all, was mine: fully furnished and provisioned, and with a stocked refrigerator and bar. Letting myself in, I walked from room to room, dropping my baggage in the bedroom, taking off my shoes, and walking in my stockinged feet into the kitchen where I opened a bottle of buttery Chardonnay. I poured myself a glass, and sat down in a huge, incredibly comfortable club chair that faced a window overlooking the building across the way and the street below.

To say I was stunned that the view from my living room window was of the very building that housed the laboratory where I did my initial fibre-optic research would be a considerable understatement. I was even more surprised the next morning when, standing out on Exhibition Road, gazing at the building where my flat was, I recognized its façade as one I'd passed hundreds of times on my way to and from work, the selfsame place I'd told myself that someday I would own.

Every day seemed to yield a new astonishment. On the day that followed the discovery that I was truly living in the flat of my dreams, I walked across the street and into the Imperial College building that housed my laboratory, only to discover a major display highlighting fibre optics research, much of it featuring my achievements at the college and subsequently in the field.

Many years later, in the summer of 2016, I took my grandson, Nikki, to London to show him the sights. Included in the tour was a stop at the Marble Arch where his grandpa had stood only a few feet away from generals, statesmen, and royalty to watch the coronation of Queen Elizabeth II, whose vast realm had once included my native land.

'I Cooked for You!'

As for my other brushes with British royalty, they all involved Prince Philip, the Duke of Edinburgh, who, when he gave up polo in his sixties, turned to the gentler, yet still highly competitive sport of carriage driving. As practiced by the Prince, this involved racing a stripped-down-though-quite-proper carriage drawn by a four-horse team and competing against similar carriages going around and around a meticulously groomed oval track.

It was alongside one such track, on a cool autumn day when the Prince was racing, that Satinder and I sat under a sizable crisp, white tent, both of us dressed to the nines as one might only be when watching British royalty at sport. Liveried staff passed silver trays bearing exotic hors d'oeuvres and crustless cucumber sandwiches along with a limitless supply of Champagne and other libations – no Campari for Satinder this time, though: too plebeian.

Descending from his carriage, the Prince gave the reins over to one of the two grooms – or were they bodyguards? or both? – who had been aboard the entire time the Prince was racing. He then made his way directly to the small white-clothed table where Satinder and I were sitting.

'Dr. Kapany, right?' He extended his hand, and then, realizing he still had on his riding gloves, he pulled off the right glove and shook my hand.

'Yes, of course,' I said, 'and my wife, Satinder.' She stood shyly as if to curtsy or bow, when the Prince pooh-poohed the formality, then shook her hand, as well.

'I want you two to join me for an informal dinner tonight…' he began, pausing for a moment as if to let the invitation sink in – 'a dinner,' he added, 'that *I'll* be cooking. So, please don't expect anything too elegant or too fussy.'

I should note that this was not the first time that I'd been in Prince Phillip's company. For a number of years I had been a member of an ecumenical religious organization that he had established to foster a greater understanding and tolerance among various religions. We'd had at least a half-dozen semiformal lunches together at Buckingham Palace as a group. I'd also made a financial contribution to the organization that the Prince insisted that we announce to the media. So, *lunch* with the Prince, yes. But I'd never had *dinner* with him. And certainly not one that he himself cooked.

A few hours later, Satinder and I arrived at the dinner venue where we were led into a good-sized drawing room to mix with a dozen other guests, and then into a quite homey dining room where the Prince stood at a resplendent portable barbecue, grilling hotdogs and hamburgers and placing them on large Spode platters to be passed around a relatively long rectangular table with chairs placed around its perimeter, the 'head' being at the table's centre. Large bowls of potato salad and coleslaw – presumably made by the Prince – sat in among a gaggle of condiments that were passed, along with frothing, amber pitchers of ale.

His cooking duties complete, the Prince took his place at the table's head, insisting that Satinder sit beside him and that I sit directly across from him. As I recall, there was no preassigned seating for the others, and in short order the conversation at the table became, if not loud, then somewhat raucous. Everyone was having a good time – particularly, it seemed, the Prince and Satinder, who were engaged in a spirited discussion about, of all things, *me*, joking and laughing about some of my husbandly quirks: at least, that's what Satinder told me on the way back to our apartment – that, and whatever I could overhear at the same time that I was trying to make scintillating conversation with the women on either side of me.

I suppose it pleased me that my quirks were so entertaining to the Prince, though I was more than a bit curious which quirks they were and whether they would provide pillow talk later that evening for the Prince and his Queen.

Some years later – I assumed the Prince would have forgotten our charming barbecue dinner – I attended a Royal Academy of Engineering banquet dinner. It was a black-tie event honouring the new Fellows, of which I was one. As the head of the organization, Prince Philip was also in attendance and it was clear that everyone was trying to chat him up, the old Fellows and the new Fellows alike. I decided I'd keep my distance before I approached him to say hello, until the crowd, hoping to catch his ear, had thinned. As it turned out, the Prince caught mine first: I was almost at the opposite end of the banquet hall when he suddenly called out, his voice penetrating the crowd and echoing through the room, 'Kapany! I know you! I *cooked* for you!'

Narinder with Prince Philip (left) at a gathering of the Royal College of Engineering, London.

Satinder and Narinder with Prince Charles.

Narinder's granddaughter Tara Kapany visits Imperial College London.

PART X

BELONGING

Facing page: Watercolour and pen on paper, artist Sumeet D. Aurora.

Belonging

A good friend of mine was recently asked to join six other men for dinner at an extremely classy restaurant. He was hoping to become a member of the exclusive club that they belonged to. And they had engineered this dinner to determine whether he was, in their words, 'a clubbable man.' He was admitted, so I guess he was – clubbable, that is.

Now, if the same question were posed to me, I would likely say no. I may be a social animal, yes. But hardly a clubbable man.

Yet, as an innovator, businessman, entrepreneur, engineer, farmer, philanthropist, and art collector, I found no shortage of organizations to belong to, where I would be a good fit, learning from others, exchanging ideas, and providing what wisdom and life experience I could of my own. If I wanted to be part of something larger, in other words, there were plenty of paths I could follow. The Royal Academy of Engineering, where my former dinner companion Prince Philip was a Senior Fellow, was but one such organization, and I was honoured to be selected as a fellow as well.

There are three other organizations in the long list of my memberships – some lifelong, others shorter term – that I feel bear special mention, and that have had more than a passing effect on me. And that, in one case, has affected me profoundly.

The first is the Young Presidents' Organization. Founded in 1950, the YPO was a network of relatively young chief executives that numbered a few hundred and was largely US-based when I joined back in the mid-1960s. In my late thirties, I was one of the organization's middle-aged members. Not only was I its sole Sikh, but also its only Indian.

Still, among this vibrant and energetic group of insiders, I never felt like an outsider. On the contrary, I was warmly welcomed and quickly made to feel at home in the organization that today, incidentally, counts over 24,000 members in 130 countries, making the YPO one of the world's largest organizations of business leaders.

Beyond the YPO's dynamic and fast-growing, by-invitation-only membership, what impressed me the most about it was its genuine commitment to lifelong learning, as manifest in the club's ongoing sponsorship of seminars, conferences, symposia, training sessions, and other events and resources, including its own burgeoning library of recorded Ted Talk–like presentations covering a smorgasbord of topics available to any member.

The fact that most of these events were staged in some of the world's most intriguing locales, cosmopolitan cities, and exotic vacation areas and spas made the whole experience that much more special. Not surprisingly, I felt most at home at those events that focused on innovation and entrepreneurship, and I was a frequent contributor during the formal presentations themselves and in many casual after-hours conversations at the bar. We were hard drivers, we YPO members, and there was many a time that we were still engaged in deep discussion or celebration when last call came around.

Priding itself on offering a rich bounty of global networking opportunities, one that only such a powerful assemblage of individuals could make possible, the YPO as an organization was – and still is – constantly trying to shake things up, in a good way, for its members. It was a few of my YPO colleagues, for example, who first recommended that I apply for a high-level post in the Nixon administration, and then worked with me to further my application.

I recall one YPO junket in particular, that Satinder and I attended together, a week-long conference ultimately in Vienna. I say 'ultimately' because somewhere in flight over the middle of the Atlantic one of the conference organizers decided that our group of a half-dozen should lay over for a day in New York for an evening of theatre, before catching an impossibly early flight to Vienna the next morning. As I was already jet-lagged from the flight from California to Europe, I don't remember much about the New York show that evening beyond the presence of the Duke and Duchess of Windsor – the brother of King George VI, and his wife, Wallis Simpson – yet another brush with British royalty!

By the time we arrived in Vienna the next morning we were all truly exhausted. Thankfully, our beds were still turned down and waiting for us from the night before. Satinder and I were already warmly tucked in when the phone rang. It was our conference organizer again. 'Put on your best duds,' he said, 'we're going to the opera to catch a matinee.' The opera house was an elegant, gilt-trimmed, wedding cake of a building. As for the opera itself, I remember even less about it than the theatre the night before, sleeping as I did from the first note to the last. And the conference in Vienna that followed our afternoon on the town, it was as rich, delicious, and deeply fulfilling as the desserts we had at Demel's: all with the ubiquitous Schlag.

A second and very exclusive organization that I have the privilege of belonging to is the Cosmos Club, a posh, understated, and intellectually challenging private club that numbers among its present and past members more than three dozen Nobel laureates, twice that number of Pulitzer Prize winners, more than a minyan of Supreme Court justices, and, last but not least, a smattering of American presidents and vice presidents.

Headquartered on Embassy Row in Washington, DC, the Cosmos Club is dedicated to the advancement of its members in science, literature, and art. To that end, the club's facilities – its library and its common rooms – were designed to encourage both the meeting of individuals and of minds, and the open and spirited encounter and exchange of ideas and belief systems. As a result, conversations tended to be far-ranging and often surprising, evoking fresh and original thought and providing the foundation for new friendships and powerful work and life experiences.

Flying in the face of these noble precepts was the club's 'men-only' policy, in effect for 110 years until 1988, when the club finally opened its membership to women. Before that year, any woman – even the spouse or guest of a member, even if she had been invited to a meal in the swank dining room, or a drink in the clubby, turn-of-the-century bar – had to both arrive and leave through a side door. I thought it absurd, and heartily applauded the change in policy. Today, not only does the club welcome women members, its president is a woman.

Aside from the conspicuous absence of women when I first joined the club, I was struck by the ages of most of my fellow members. While age-wise I pretty much fit in with my fellow YPO members, at forty, I was the youngest of my Cosmos Club brothers, most of whom appeared to be in their sixties and older – considerably older. Sitting in the Cosmos Club lounge was like sitting with Father in the parlour all over again. Which was good – terrific, in fact. Only now, these grey hairs, these noble personages were in some ways my peers. It boggled my mind.

Still, the club that had its most lasting effect on me was not the YPO, with its cohort of CEOs and youngish doers and leaders. Nor was it the Cosmos Club, with its mature and elegant solons of science and creators of literature and art with whom I could engage virtually any day. Rather, the organization that spoke to me more than any other was the National Inventors Council (NIC), a scintillating group of many of the nation's most credible, original, and prescient inventors. Initially the brainchild of President Franklin D. Roosevelt, meant to foster invention in support of the war effort, the club survived his passing, the end of the war, and four presidential administrations until it succumbed to President Nixon's disdain for eggheads and geeks –

though a phantom, private-sector version of the Council managed to stay in business for a number of years, during which time its mission evolved to helping other often-lone inventors bring their innovations to market and navigate the rather forbidding shoals of the patent process.

As best I can recall, there were some twenty of us who were active in the mid-1960s to the late '70s when I was a member. We'd meet three or four times a year at various locations to connect up with other individuals – sometimes not inventors – whom we could learn from or teach or, most likely, both, the essence of which we could later bring to our own work. There were great but low-key collaborations that fostered thinking that often brought the inventors' process to a wide range of fields and provided new insights for our group.

I remember one visit to Washington, DC, for example, when we spent the better part of the day with then–Vice President Hubert Humphrey. Following an informal and inspiring talk by him, we literally rolled up our sleeves and shared our ideas about the power of inventions to create a better future. There was no question that Humphrey 'got it' and that he came away from our time together compelled by the process of invention and well-informed about the best means to apply it in the global arena.

A few years later, Council members travelled to Little Rock, Arkansas, to advise a young politician who had just made a successful run for the governorship, making him the youngest governor in the United States.

'Look,' Governor Bill Clinton said to our little contingent, 'I've made commitments to the people in my state to bring high technology to Arkansas, and I want you to help me.'

What struck me about Clinton was what a quick and thorough study he was. He had an inventor's aptitude: the ability to follow an idea, no matter how complex or arcane, to wherever you were trying to take it. Better yet, he had the intelligence and facility of language to reframe that idea so that it made even greater sense.

It was in 1972, a decade before I met Clinton, that I attended the NIC event that set me to thinking anew about my life, not in a particularly earth-shaking way, but still with a profoundly different personal perspective. And it would take a trip with a bunch of other inventors to the Space Center at Cape Kennedy on 16 April that year to witness the launch of Apollo 16 – the tenth manned mission of the Apollo program, and the fifth and second-to-last to land on the moon – to bring that new perspective into focus.

A little after noon, the rocket carrying the Apollo craft blasted into a nearly cloudless, brilliant blue spring sky, burning liquid oxygen and liquid hydrogen propellant at a furious rate, creating 7.7 million pounds of thrust as the rocket lifted off the launchpad and cleared the tower. On VIP grandstands and on wide swaths of lawn a safe distance away from the take-off, onlookers, many in shirt sleeves and carrying cameras or binoculars, felt the earth vibrate, heard the thruster engines roar, and saw the sky part.

As did I. But my thoughts during these first few seconds of flight were not of the moon as a destination or a chamber for experimentation. My thoughts weren't even of manned flight. They were of the moon I'd studied some forty years ago from the floor of an oxcart drawn by a lumbering pair of bullocks. And if I wondered back then about the moon, it was only to see the reputed man *in* it, not *on* it. That was my universe, and I was of it. More wondrous than curious. A child of *that* time, of the 1920s, and under *that* shining, undisturbed moon.

But now, after a few hours spent at Cape Kennedy in 1972, still only distanced from that rocket by a few thousand miles as it was about to change its trajectory and go into Earth orbit before making for its lunar destination, I suddenly thought of the moon anew, no longer from the perspective of the last generation, nor even from the present. But from the *next*. The inventors' next. The sky *is* the limit, I suddenly, almost begrudgingly, yielded, here on this spit of land where I could both see and smell the contrails of the mother ship, everything up close and personal. We were, all of us, on the cusp of a different world. A time of fantastic change. The most exciting generation. The most dangerous generation… with the power to preserve humanity and the power to destroy it.

No oxcart, the rocket was moving at warp speed. As were the insights that would be gleaned from it. Where will we go? What will we learn when we get there? Where will *I* go? What will *I* learn when I get there?

I lost some of that boy and his tightly subscribed, safer life on that day in 1972. I became less of a child of the past that afternoon in Florida, more of a man driven by the spirit of the future. There were still spectacular things to be done. And there was still plenty of time for me, and others, to do them.

The Young Presidents' Organization

newsletter

Published twice monthly for members of the Young Presidents' Organization, 375 Park Avenue, New York, N.Y. 10022

VOL. 1, NO. 9, MAR. 23, 1971

Narinder Kapany Named YPO Ambassador to India

International Vice President Steve Yeonas has announced the appointment of Dr. Narinder Kapany as YPO Ambassador to India . . . Narinder is president and director of research at Optics Technology, Inc., Palo Alto, Calif. . . . In the past, he has had a profound interest in expanding YPO programs in India. As Ambassador, Narinder will stimulate Indian interest in YPO activities, and provide leadership in gaining new members and forming chapters . . . YPOers having helpful ideas, or any business or social contacts in India that may be beneficial to Narinder in this program should contact him at Optics Technology, Inc., 901 California Avenue, Palo Alto, Calif. 94304.

Top to bottom: Meeting the Pope in Rome; a gathering of members of the Young Presidents Organization, which named Narinder its Ambassador to India in 1971.

Narinder S. Kapany
With best wishes
Hubert H Humphrey

Vice President Hubert H. Humphrey, with Narinder and other members of the National Inventors Council, Florida.

PART XI

MASSACRE

Massacre

1984 was a year of significant disquiet and danger for Sikhs in India, particularly those who were importuning the government in Delhi for a more equitable allocation of water resources and to make Punjabi the official language of North India. In support of these requests and others, there were a number of incidences of picketing around the country, almost all of them peaceful. Among the most vocal and visible of the protesters was a cadre of Sikh separatists who, according to the government of India, was stockpiling weapons.

Back in the United States, a number of prominent American Sikhs staged a teach-in in Washington, DC, in March to educate the US Congress as well as to update the widespread American-Sikh populace about the situation in India. Some 30 or 40 senators and congressional representatives were in attendance, along with more than 400 Sikhs from around the United States and Canada. And I was invited to address them.

A powerful voice on the Sikhs' behalf in this group was that of my own congressmen from California, Pete McCloskey and James Corman. Warm, earnest, and forthright, McCloskey first won my trust and, subsequently, my abiding affection. Over the years he became a true and genuine friend, not only to me personally, but also to the growing American-Sikh community that was rapidly becoming part of the nation's diverse population – from its cities to its rural areas and from its high tech communities to its college campuses.

In my keynote to the teach-in, I told the American-Sikh story, emphasizing the similarities between American and Sikh belief systems, particularly the shared concepts that all men and women are created equal, and that everyone should be given the opportunity to rise in station commensurate with their abilities and ambitions.

I also spoke of the then–highly charged and incendiary political situation in India, and of my fears about how the Sikhs in that country would fare if the Indian Army were to attack their Golden Temple in Amritsar. Such an attack, I ventured, would likely signal the beginning of a government-fuelled Sikh genocide.

Not long after the teach-in in Washington, I found myself increasingly responding to well-supported rumours and scuttlebutt emerging from India – that the government was indeed about to attack the Golden Temple on the pretense that it was being used by insurgents both as a major storehouse of arms and as a safe house for Sikh 'terrorists,' as the Delhi government had recently designated a number of the separatists. With the hope of helping to defuse the situation, I sent a telegram to the Prime Minister, Indira Gandhi. In it, I suggested that a number of American-Sikh leaders travel to India to sit down with Indian Sikh leaders to talk the situation through. I never received a response.

I should mention that this was not my first experience with the Prime Minister, and that our prior meeting, while quite innocent, made me wary of her. It had taken place in the late 1960s, during the Prime Minister's first term, when she was bestowing awards in the form of medals on a number of Indian scientists who had distinguished themselves abroad in their respective fields. I was the only turban-wearing Sikh among the recipients, and when Mrs. Gandhi stepped in front of me to place the sash holding the medal around my neck, I bent forward in her direction to make it easier for her to manoeuvre it over my turban. Still, the sash got caught up on the turban and the medal ended up dangling in front of my face. There was some laughter from the audience – nervous laughter, actually, as I quickly offered, 'Here, let me have a go at it.'

What followed was the briefest transaction, as the Prime Minister removed her hands from my sash and I replaced them with my own, a movement that put us face-to-face, only inches from each other. I remember that I smiled, thinking the moment both charming and humorous. But her expression – embarrassment followed by a flicker of anger and resentment – suggested otherwise. And from that moment on, I ceased to fully appreciate Mrs. Gandhi.

Flash forward to 1984, when throughout India anti-Sikh sentiment was growing, in the capital, in the countryside, and also abroad. There were hints of it even in the cosmopolitan San Francisco Bay Area. For example, I'd recently hired a young Sikh woman to work for the Sikh Foundation in Palo Alto. She had previously worked with the Indian consulate in San Francisco, which, she reported, had hired a fellow to find out as many details about the American-Sikh community as possible. Subjected to his interrogation, she confirmed to knowing a Dr. Kapany.

'You know Dr. Kapany?' The inquisitor sounded seriously concerned, the young woman recounted to me. When she told him, 'Yes,' he quickly responded, 'Well, you'd better be very, very careful of what you say around that man! Because when Kapany talks, America listens.' America, perhaps. But definitely not the powers that be in Delhi.

For on 6 June, 1984, a time of a major gathering of Sikhs to mark the martyrdom of Guru Arjan Dev, our sixth Guru, Indian Army troops were ordered to attack the Golden Temple, the most holy Sikh place of meeting and worship. Despite the valiant resistance of those praying or visiting or even residing there – including a retired Indian Army general who had come to help stave off the attack – the Golden Temple was overrun in a matter of days. More than 3,000 souls were massacred, including hundreds of women and children.

To say that I and tens of thousands of others in the United States and around the world were angry would be an enormous understatement. Within a few days, we – the American-Sikh leaders and the congressmen and women who'd supported us in the past – proposed that we go as a group to Amritsar to gather data about the massacre and then return to the United States to report on the tragic events. It was hardly a surprise when my request for a visa was denied. 'What you are seeking to do,' the Indian ambassador to the United States wrote to me, 'is an unwarranted intrusion into Indian internal affairs.' No visa for me, in other words. Nor for any of the others in our group.

Still, we persisted in the belief that the world should know about the massacre, that it shouldn't become just another barely acknowledged stain on India's history. To bring what happened to light, the Sikh Foundation raised the money needed to place informative, full-page notices about the massacre in major newspapers in New York, Washington, Los Angeles, Chicago, and San Francisco.

The Foundation also arranged speaking tours to Sikh temples and other places of worship as well as to radio and TV stations and schools throughout North America. One of these events, in Vancouver, British Columbia, was attended by more than 3,000 guests, including a number of Canadian government officials. It was the first time that many in the crowd had ever heard of the massacre. Many, in fact, knew very little about Sikhs at all, and even less about the ongoing strife in our home country.

Then, on 31 October, 1984, less than five months after the massacre, Prime Minister Gandhi was assassinated by two of her Sikh bodyguards in Delhi.

Within days of the assassination, Sikh homes throughout India were besieged by drunken mobs of prisoners released from jail, it is generally believed, with the support of government officials and local police precisely for this occasion. Giving credence to this belief is the fact that only Sikh homes fell to the rioters, and only Sikhs were killed, with the death toll rising to more than 3,000 in Delhi alone.

By the end of 1984, some 6,000 Sikhs had succumbed to violence and prejudice in India. Yet in all this time – from then to now – not a single person has been sent to prison, neither the killers nor those who gave the order. On the contrary, a number of major government officials from back then remain in their positions today.

Top: Darbar Sahib, late eighteenth century, paint on ivory, 33 × 22 cm, Kapany Collection.
Bottom: The Golden Temple of Amritsar, Kapur Singh, 1886, oil on canvas,
75.6 × 52.8 cm, Kapany Collection.

The Sikhs of India are much like the American colonists of 1776.

Why should Americans care...

Because we are Americans

The Sikhs: An Admirable People with Progressive Beliefs

The Massacre: Outrage at the Sikhs' Holiest Temple

The Aftermath: Grief, Anger and Continuing Repression

Righting the Wrongs: Some Immediate Goals

What you can do to help

੧੬

THE SIKH TIMES

Vol. 2 No. 23 October 12, 1984 ECHO OF THE SIKH WORLD 40¢ (Anual Subscription $20)

INDRA FINDS EXCUSE TO EXTEND PRESIDENT RULE

SIKHS MADE SCAPEGOATS FOR POLICE BRUTE KILLINGS IN PUNJAB

For Information

Palo Alto-

The following agreement has been made between THE SIKH FOUNDATION U.S.A. and SURAJ CORPORATION:- Wher as, The SURAJ CORPORATION owns the publication called SURAJ and THE SIKH FONDATION has under taken the publicatio-n of THE SIKH TIMES Now, therefore the parties agree as fo-llows:-
1. THE SIKH TIMES and THE SURAJ shall not merge.
2. THE SURAJ CORPORATION will own SURAJ.

New Delhi: October 5, 1984- Indira has played yet another political gimmick to justify the extension of President Rule in Punjab by six months more to last till April 5, 1985. She clearly corelated it to an incident in which a hand-grenade attack was made by some unknown miscrents at a Hindu religious festival of Susserah at Mansa Mandi in Punjab, falsely accusing the Sikhs.

Our special correspondent, who visited Mansa Mandi to probe deeper in this incident, informs us on the phone that even the local leaders do not rule out the possibility of backstage role of police having arranged the throwing of the hand-grenade.

It is on record that the Indian Parliament mended the Constitution two months back allowing Gandhi to rule the state for another one year. How she could realise two months earlier that an attack would be made by the Sikhs on a religious function on this particular day. This is a camouflage to arouse the communal passion to give the Sikhs bad name and per-etuate her direct rule by another year. The President Rule was imposed on Punjab on October 5, 1983 which was to lapse on October 4, 1984.

Clockwise, from top left: Promotional advertisement about Sikhs published in US newspapers, 1984; Narinder and other Sikhs confer with members of the US Congress in Washington, DC, 1984; California Congressman Pete McCloskey; front page of the Newspaper *Sikh Times*, published by the Sikh Foundation in 1984–1985.

PART XII

FAMILY MATTERS

Facing page: Watercolour and pen on paper, artist Sumeet D. Aurora.

Once-in-a-Lifetime

This is the situation: there are six of us – along with a driver doubling as a guide – in each of the two open 4 x 4 safari vehicles, Jeeps, I think, maybe Land Rovers. Satinder, Kiki, and I are in the lead vehicle. Kiki's husband, Michael, and their two kids – our grandchildren – Ari and Misha, are bringing up the rear in the second vehicle.

It is still early in the day, before seven, and the guide speaks softly, just loud enough to be heard in our vehicle, tooling along the gently rolling hills that surround the huge, luxurious Phinda Getty House where we are all staying. I look back to the guide in the Jeep close behind us to see if he is also alerting his passengers, likely telling them the same thing: 'When we get right alongside the lions, keep very quiet. Don't make any sudden movements or any loud noises. They're used to the Jeeps; they think they're animals that won't harm them.' Then, with just the slightest edge to his voice, he repeats, 'No sudden movements, no loud noises.' Off in the distance, I already hear the cacophonous roar of lions. Or at least I imagine it.

I'm sitting in the second row of seats behind Satinder as we make our way closer to what appears to be a pride of five lions taking in the early morning sun in the private Big Five Wildlife Reserve where the Phinda Getty House is situated. I glance at my wife, then look through the windshield to the lions now up ahead and about 100 yards from us, the Jeep just bumping casually along.

I look directly ahead of me at Satinder, sitting next to the guide. She is pressed up against the low steel door – no window, no window frame, clearly built for observation, not protection – 'riding shotgun' but without a weapon. She is fast asleep, having nodded off to her right, in the direction of the great outdoors. We are now only a few seconds away from the nearest lion, a somewhat bedraggled male – though still a lion, still a 'big cat,' still regal-looking and aggressive enough to command a pride. And with my wife, my lovely Satinder, now only a few feet away. Lazily, he opens his mouth as if to roar, though he doesn't. I can see his teeth. Is Satinder going to open her eyes and startle? Scream? Clamber back inward into the Jeep and into my 'protective' albeit, more realistically, useless arms? Useless, that is, to fend off a 400-pound lion that finally realized that a Jeep is *not* just another friendly animal. I pray Satinder will continue to sleep. It is too late to wake her now…

It was to be our best family trip ever, a 'once-in-a-lifetime experience.' At least that's what Kiki promised when she first introduced the idea of a South African wild game safari about ten years ago. 'Better than Rio. Better than Japan. Better than China,' she insisted, ticking off a number of prior Kapany family adventures.

To reach the Big Five Game Preserve from London, we first flew to Capetown for a few days' visit, then to Johannesburg, and finally by private plane to the preserve and the Phinda Getty House where, for a mere $1,000 per-day-*per-person*, we were its sole occupants – along with our own ranger/host, a butler, a chef, and a constantly changing cast of servants who all managed to appear only when help was needed.

The house – a lodge, actually – was nothing short of magnificent, with four enormous bedroom suites and literally thousands of square feet of common area. Stylish and sophisticated, set in total seclusion among lush, rolling green hills with spectacular vistas from every suite and even from every bath, the lodge was a showpiece of thatch and timber, sandstone and screed concrete, with incredibly tall, vaulted, and beamed ceilings and the most tasteful and comfortable, human-scale furniture.

The house was also ideally located for walks through the bush and for wild game drives. There was even a special white rhino tour that we opted out of for fear it would take too much time from the precious seven days we had booked. As it was, the six of us set out for the bush in the two safari vehicles twice daily – at 6:30 in the morning, and again in the late afternoon – for an average of three hours each time out. While that may seem like a long stretch, it wasn't. It was usually over before we knew it. Besides, there was always some wildlife adventure to hold our interest: a pride of lions taking down and then feeding on a zebra grazing at the edge of the herd, or two other beasts locking horns and fighting for social supremacy.

Whether it was dumb luck or Kiki's genius scheduling skills, our seven days in South Africa coincided with both the final week of tennis at Wimbledon and the World Cup soccer finals. Although the lodge's owners advertised it as a no phone/no Internet/no TV retreat, they soon acquiesced to a TV in each of the suites and a stadium scoreboard–sized flat-screen television in the enormous living room, where, returning from the bush, we'd sit back and watch the tennis or the soccer, then later enjoy a terrific meal the chef prepared. Either that or simply sit on the deck and take in yet another magnificent red-and-orange-painted sunset. Best of all, though, were those moments after the sun had gone down and the torches were lit, leaving the six of us usually in the same room to reflect on the day's events and wild-animal sightings, laughing and talking and being a family.

The Jeep that Satinder, Kiki, and I are in slows, then stops. The driver stands. We are no more than 20 feet – Satinder the closest – from the nearest lion, the bedraggled male. He lazes indolently in the sun, calm, though perhaps deceptively so. I study the back of Satinder's neck, waiting for her to startle awake and cause the lot of us to be taken down and be eaten alive like that poor zebra. I try to alert the guide to the situation, but, standing in the Jeep right next to her, he seems unfazed, his rifle lying impotently on the vehicle's floor.

Just then, Satinder does seem to rouse a bit. But no, she falls back to sleep again. I still remember clearly seeing her in profile, her mouth slightly agape as she dreamt sweetly, blissfully unaware that were she to open her eyes, she'd be face to face with a lion – a scraggly male, but a lion nonetheless.

'Oh, my goodness!' I imagined her saying softly, reverently, and barely above a whisper if she were wake. 'Oh, my goodness!'

The family enjoying an African safari trip.

Kids

My greatest and most rewarding invention, by far – mine and Satinder's, that is – was and continues to be our children, Raj and Kiki, and our grandchildren, Ari, Misha, Tara, and Nikki.

When Raj was eight years old, we sent him off to Eton in the UK. Well, not exactly. We sent him off to the favoured *pre*-Eton school where success there might mean an easier go of it for him at the actual Eton. The problem was that, while Satinder and I may still have been Anglicized from our earlier grand adventures in London, Raj never was. Instead of loving, or at least acclimating to merrie olde England, Raj hated both the British weather (cold) and the British food (lousy). His situation in England was, in fact, the exact opposite of his in California, where he and his pals played in the long, warm, lazy afternoons and enjoyed home-cooked meals, or chowed down a few days a week on fast food.

'For breakfast,' he wrote grimly in an early letter from school, 'we get over-ripe tomatoes that everyone hates. No one ever eats them. Instead, we just smash them with our spoons and send the seeds flying at the others across the table.' Even Satinder's moving into a nearby flat for a few weeks to tend to Raj's wants was of no consolation.

We did all take notice, however, when, in his first visit home after six months in England, he called me 'sir.' We were sitting in the car at a stoplight when it happened. Satinder nodded, as if this were some great developments that we should all take pride in. I smiled, the smile of a satisfied (for-the-time-being at least) father. And Kiki, Raj's younger sister by two years? She couldn't stop laughing. 'Did you hear that?' she exclaimed. '"Sir!" Can you believe it?'

Actually, with Raj away at pre-Eton, and her mother a full continent and ocean away tending to him, Kiki had me all to herself, at her full beck and call. And don't think she didn't know it. Years later, she would explain to me that we were both Scorpios and were therefore both in need of good wives. And that with her mother temporarily away in England, Kiki had impressed me into service. *Her* service.

I wasn't much of a cook, though, nor did I have the time or the inclination to prepare proper meals. As a result, we tended to go out almost every night, and usually to a restaurant of Kiki's choosing. She had two favourites: McDonald's and an extremely hoity-toity, white tablecloth, French supper house called L'Auberge in nearby, enormously well-to-do Atherton, the next town over from semirural, woodsy Woodside, where we lived then.

I still remember sitting across from her in the largely candlelit room, the huge, leather-bound menu propped up in front of her, her head occasionally popping up over the *carte* for a few seconds with questions like, 'What's an escargot?' or 'Would I like sweetbreads?' She tried them all, was an adventurous eater, and it wasn't long before the fare at McDonald's no longer captured her interest, even just to go slumming. Besides, the L'Auberge staff and management were on a first-name basis with her where, at McDonald's, they didn't know Kiki from any other quarter-pounder-consuming kid.

Still, all good things had to come to an end, and two years into pre-Eton, Satinder and I relented and Raj moved back home. Not long after, 'Auberge,' as Kiki and I had come to call it, went out of business, quite possibly, we decided, for lack of our patronage.

I'm not sure if it was revenge for all those rank breakfast tomatoes and reputedly thin gruel that pre-Eton served its up-and-coming twits, but once safely re-ensconced in his familial home, Raj managed to get himself into a good bit of not-bad trouble.

Pouring water into my car's gas tank to, as he claimed, (a) save me on gas, and (b) just to see if it would run on it, may have been mere curiosity or a devilish acting out.

And then there was the night when a friend of Raj's, Mark Curtis by name, was sleeping over and the two of them climbed out his bedroom window and walked a good mile to the house of a third friend to commit who-knows-what mayhem. And they probably would have been successful at it had the eagle-eyed small-town police not spotted Raj and his friend tossing pebbles at their other pal's window to rouse him from bed.

The two teen perpetrators appeared at our front door accompanied by two officers. 'Is one of these yours?' the taller officer asked. 'Yes,' I said.

Raj and Mark were definitely scared, shaking. Over their shoulders I could see the red light on the top of the squad car sweep the garden every few seconds, illuminating my face with every pass. After the taller of the two officers admonished Raj and Mark a final time, I took both boys into the house. I supposed I had to punish Raj, of course, though my heart wasn't in it. Also, it seemed unfair to punish Raj and not punish Mark, which I didn't believe was in my purview. Nor was I about to rat out Mark to his father. So, all in all, the two boys pretty much got away with their evening's misadventure. And while my inaction was certainly not a feather in my parenting cap, I was pleased to come out of the entire event a relatively good guy.

Some years later, when Raj was in high school, I wouldn't let him play football for fear he'd sustain a concussion, only to have him take up wrestling in college, at the University of California, Berkeley. As for Raj's academic career, it was as much a mystery to me as mine was to my parents. I never once saw him crack a book on the frequent weekends he would spend at home with Satinder, Kiki, and me. And the answer to any questions we might have had about his progress in school was always 'fine,' which, we learned from other parents of teenage boys, was the answer their sons *all* gave, the answer echoed in dozens of languages around the world.

As it turned out, 'fine' was good enough, and after a stint following graduation where Raj worked for me at Kaptron, and subsequently in the London office for AMP, the company that bought Kaptron, he went out on his own and started a number of successful companies.

In 1994, Raj married Erica and they raised two children, Tara and Nikki, before they divorced in 2008. Tara finished college in 2017 and is now working in the real estate industry in San Francisco, while Nikki graduated from college this year and will start his job in a new entrepreneurial company.

If Satinder and I never saw Raj study, our experience with Kiki was the exact opposite. Kiki, it seemed, studied all the time, feverishly devouring one book after another, writing one, two, three papers at a time. She also had an affinity for languages, winning an award for her work in high school French class while also becoming proficient in conversational Spanish. Her dedication to her high-school assignments was so great that sometimes I'd even tease her about it – as one over-achiever to another – jangling the not-too-private, casbah-like beaded curtain entry to her room and sticking my head through the beads to ask if she was almost through.

When it came time for college, Kiki followed in her brother's footsteps, attending UC Berkeley, about 45 miles from our home. Every few weeks Satinder and I would drive up and take her out for dinner. And no sooner did she finish her last bite than she excused herself to return to her beloved books. Thinking back, I may have teased her for her studiousness, but I was enormously proud of the depth of her commitment to learning, from high school to her time as an undergraduate, and later, in law school.

Unlike Raj, who is more business oriented and likes things done his way, Kiki has a greater instinct for family and tradition, recently arranging a reunion in India for some fifteen close relatives. It's not that Raj doesn't share that instinct; he definitely keeps in close contact with a few relations. It's just that Kiki is more inclusive and collaborative, keeping alive the Kapany family culture and nurturing strong family relationships.

In 1990 Kiki married Michael Schwarz, an award-winning documentary filmmaker, in a Sikh wedding ceremony that was worthy of a documentary all its own. Jewish by birth and upbringing, Michael, nevertheless, donned the turban and ceremonial embroidered long shirt, sash, and trousers of a Sikh bridegroom for his wedding. True to Sikh tradition – as interpreted by two very old Sikh friends of mine from the 1950s, Professor Janmeja Singh and Dr. Gurnam Brard – Michael rode up the driveway to the bride's, i.e., Kiki's, family house on horseback, followed by his groomsmen, also in turbans and celebrative Sikh wedding garb.

At the gate, Michael dismounted and walked with the others the rest of the way to the house, where he met up with Kiki before an elegant altar we'd had erected in the great-room where the actual ceremony was to take place. At its centre, the Sikh Holy Book lay open. No chairs, though; guests, except those whose physical condition prevented it, were to sit on the carpet.

Without exaggeration, Kiki's entrance took my breath away. She looked absolutely radiant, dressed in a traditional Sikh wedding dress. I don't think that she ever looked more beautiful. Or more like her mother.

A Sikh priest led the opening prayers as the guests prepared for this holy union of two souls, this partnership of two equals. Following the intoning of these initial prayers, formal introductions of the wedding party were made, followed by an elaborate breakfast and later a luncheon. The sacred verses from the Holy Book, the Guru Granth Sahib, were then sung as I, the father of the bride, took one end of the long sash from Michael's shoulder and handed it to Kiki, thus symbolically joining them, one end of the sash now in Michael's hand, the other in Kiki's.

Then, holding the sash, Kiki and Michael walked four times around the Holy Book, interspersed with joyous singing and prayer. At the final such walk around the Holy Book, the guests, all given handfuls of flower petals, threw them at the newly married couple, blessing them with a final prayer, the scent of fresh flowers filling the room.

A party for 300 or so guests at another, even more scenic venue across San Francisco Bay in Berkeley was a great and fun occasion, big and boisterous, with lots of toasts, food, drink, and dancing.

The fruits of that marriage have been plenty over the years, including two more terrific grandchildren, Ari and Misha.

Clockwise, from top: Nikki, Ariana, Michael, Kiki, Misha, and Tara at the Sikh Foundation 50th Anniversary Gala; Michael and Kiki's wedding ceremony, 1990; Grandkids working in the Golden Temple's kitchen; (Back row) Kiki, Erica, and Raj, (front row) Ariana, Misha, and Tara.

Family Ties

The happiest times? For us as a family, that is? Without question, our numerous, boisterous family reunions in Dehradun, particularly the ones at the new house, the one that I built for my parents years back, large enough to accommodate us all, often for as long as four weeks and sometimes even more.

Our former house, the rented one we were living in when the mob came to take Gulu, was still standing on a neighbouring piece of land. But Father wanted us *each* – my three brothers and my two living sisters – to have a house of our own. So, he bought the adjoining acreage, subdivided it, and built a house for everyone but me – to whom, instead, he gave the most prized piece of land, with magnificent views of the snowy Himalayas in the distance – figuring, I guess, that if I wanted a house, I'd build it to my own and Satinder's specifications. Also, that I'd get my Americanized rear end back to Dehradun.

Meanwhile, as their children luxuriated in their new digs, Father and Mother continued to live in the old house. Even though it was what Father claimed he wanted, it didn't seem fair to me that all my brothers and sisters now had fancy new houses and that Father and Mother didn't.

So, on one of my frequent trips to India, I announced to Father, 'I'm finally going to do it!' Father didn't seem to know what I was talking about. I told him. 'I'm going to build a grand and wonderful house on my land! And,' I went on, 'you and Mother are going to live in it!'

'You're ignoring the spirit of the gift,' Father demurred. 'The land is for you, for your house. To build and live in.'

I repeated my offer. 'I am,' I insisted, 'going to build you a house. And you and Mother are going to live in it!'

And I did. I built them a lovely and sizable house with enough rooms to comfortably accommodate the entire family, and servants as well.

'It's yours,' I said to Father and Mother, handing them the keys on their first official tour of the completed house.

'We can't accept this, Bino,' Father said, handing the keys back to me. 'It is too generous.'

'I built it for you,' I told him.

'I know, but your mother would...'

And that's when I gave him the ultimatum: 'If you don't move in, I'm going to sell the house!' It was my ace in the hole. I knew that they had some trepidations about moving from their old residence – who wouldn't? – but that they would have even more misgivings about my selling the place. So they moved in. And as far as I can tell, they never regretted it. Besides, if they got lonely for the old place, it remained unrented and was only a short walk away.

Better yet, when it came time for family reunions, the new house, even with its modern kitchen and its contemporary bedrooms and bathrooms, felt like an old shoe. In the early to mid-1960s through the '80s, in particular, we had just the best parties. Satinder and I would arrive from California and the others – brothers and sisters, cousins and grandchildren, nieces and nephews, all manner of in-laws – from next door or wherever in the world they were living at the moment, and we'd all rendezvous at the new house with my parents. A quick trip to the market with a couple of my cousins and we'd return with baskets of fresh fruit and vegetables and all manner of provisions, even to include those of the alcoholic variety. And as our families grew, there'd often be twenty or thirty or more of us together celebrating the good life.

Sobering were the trips that Satinder and I took to my mother's ancestral home, the place where I first learned the meaning of the word 'property.' By the 1960s, the stately house had fallen into the possession of my mother's two brothers, Gurbaksh and Balbir, who, ne'er-do-wells in their youth, didn't grow out of it in their middle age.

I remember back when I was about twelve years old, overhearing my grandmother ask my mother late one night where Balbir was, only to discover him myself the next morning when he arrived home, still drunk, on a camel being led by a Sikh farmer. As a metaphor for my uncles' wasted lives, the house fell into disrepair. And when its weakened foundation could no longer support five stories, the uncles took it down to four, and then three, then two, and finally the one that remains today.

Visiting my parents, following a stopover at the ever-diminishing dwelling, Mother dutifully asked me about the uncles. But it was the house she was really interested in. And I still remember how disappointed she'd be when I gave her my report. To put the best possible gloss on a bad situation, she'd often tell me stories of my father's father, Nagina Singh, author of *The Law of Confessions* and a highly respected session judge in Punjab. And how, on the days he had to dole out punishment for the most severe crimes, he was unusually pensive and quiet. 'So, you see, Bino,' she'd say, 'you are made of good stuff, too.'

Mother died on 31 October, 1984, the same day that Indira Gandhi was assassinated and a curfew was imposed throughout India. I was in the United States at the time of the assassination, working with Sikh groups and the American Congress to investigate and censure the Indian government for the Golden Temple massacre. As a result of my activism, I wasn't allowed back into the country for my mother's funeral, curfew or not.

An influential member of the community by dint of her own good works and those of my father, because of the curfew my mother was not afforded the ceremonial remembrance she deserved. Thousands of people who might ordinarily have come to say their farewells were prevented from doing so. Instead, my brother Jat had to arrange for an anonymous truck to pick up my mother's body and carry it and a few close relatives to the crematorium for her cremation.

Her doctor claimed that she died of a stroke, but I think she was simply tired of life, and died primarily of old age.

Outliving Mother by seven years, Father passed at ninety-four, on 6 December, 1991. Satinder and I were in Tokyo at the time, with India as our next scheduled stop. It was early in the morning when Jat called with the news. Oddly, I don't remember a word of what he said, nor what I said to Satinder. But I do remember that I had a breakfast appointment at the New Otani, and that I went down – in retrospect, most likely in shock – to tell the man that I was to meet with that I had just lost my father. And it wasn't until then, saying it to a near-total stranger, that it hit me: Father was gone.

One of Father's favourite stories was about when he was in the British Royal Air Force on the Russian front and how he, a *mere* war photographer, would bet with the swashbuckling RAF pilots that he could eat more hard-boiled eggs in one sitting than they could. I was about ten years old when he first told me the story, back when the number of eggs he had to eat to beat all the pilots was only ten. But as the years went by and the actual contest receded into the past, the number of eggs Father could eat in the story didn't seem sufficiently prodigious to contemporary audiences (i.e., grandchildren). And so the number of eggs grew, at first by one or two eggs, and then by five and ten until it peaked at fifty, coincidentally the same number of eggs Paul Newman managed to consume in the 1967 movie *Cool Hand Luke*.

But, exaggerated or not, Father's story always elicited admiration and wonder. As it should be, for he was an admirable and wonderful man.

So too was my dear older brother, Jat, who, like me, was cured of the fever we'd both contracted in the early 1930s to live a long, fulfilled life. An exceedingly generous man, he was totally dedicated to helping others, particularly when those others were family members. Whenever Satinder and I would come for a visit to India, he'd drop everything to spend time with us. When Kiki graduated from high school and was traveling through India with a girlfriend, Jat had a car waiting for them at the Delhi airport and then drove them wherever they wanted to go – not simply one day, but every day for their entire two-week visit. And though he went through some tough financial times, Jat recovered, leaving his two sons a good deal of money. Giving: It was what he was best at.

Unfortunately, what he was least good at was taking care of himself. He never visited a dentist and rarely went to the doctor. I last saw him about three months before he died. He was ill, deteriorating, and I was aware that it likely was the last time I'd see him. He was bedridden, sometimes able to talk and make sense, other times unable to string together a cohesive sentence. When I left him there in that bed, I felt like I'd lost my best male friend. And it's true, I had. He was a great guy. Very honest. Very fair. I loved him a great deal.

My younger brother, Gurdev Singh, eight years younger than I, is my only surviving sibling. He was only a kid when I left India, so, unlike Jat, I barely got to know him. Ironically, he lives closer to me than any of my other siblings, by thousands of miles – in Los Angeles, a retired high school math teacher.

A happy-go-lucky guy, he was most recently going through a tough time, helping his wife through her condition of Parkinson's disease as best he could. But after a while, it got to be too much for him – he was having difficulty moving her from the bed to the bathroom, for example – so he had to put her into assisted living. He lives alone now. So do I. I should visit him.

The Kapany family get together in India, 2004.

Passing

Everything has a beginning.
And in the case of the disease that ultimately claimed my precious Satinder, its beginning, I believe, was triggered some fifteen years before the appearance of her first symptoms by the inhalation of poisonous fumes from the chemicals she used to fertilize our abundant garden – nearly two acres planted in fruit trees, flowers, bulbs, shrubs, vines, and lawns, all thriving through fertilization. And Satinder provided it, hauling open bags of the stuff – with the help of a gardener – in a wheelbarrow, from plant to plant, plot to plot. Sometimes, she'd work in a light mist and the fertilizer would become moist, its fumes even more virulent, or so I believed, occasionally inadvertently inhaling them myself, the chemicals singeing my nostrils as I stood on the periphery.

'Don't *do* this!' I insisted one morning more than three decades ago, and after that dozens of times more. 'These fumes are going to give you some terrible problem,' I said, as certain of it as I was of science itself. 'You'll see.'

She stood before me in her gardening outfit: jeans, a plaid flannel shirt, a peaked straw hat, the drawstring around her neck, suede gloves with gauntlets, and a trowel in one hand. That morning, too, she wore a bandana that covered her mouth and nose. 'I couldn't be better protected,' she said, her voice slightly muffled by the bandana, which I knew she donned only to put my mind at ease.

I wasn't at ease, however. Whatever government agency sanctioned the use of fertilizer by middle-aged Indian women, or anyone for that matter, didn't know what they were doing. And incidentally, wasn't it fertilizer that also found its way into the making of bombs?

'Don't do it, Satinder,' I implored, but to no avail.

Years later, when Satinder was in the throes of late-stage Parkinson's, I would share my opinions on the dangers of inhaling fertilizer fumes with Satinder's doctors. I'm not certain if they were humouring me, but they all seemed to agree that, possibly at least, the fumes could have helped trigger the disease.

But wait. I'm getting ahead of myself. It was at least fifteen years after expressing to Satinder my concern about the fertilizers that we had the first inkling of the terrible problem that they might have precipitated. In 1990, Satinder and I were vacationing in Hawaii, at the Mauna Kea Hotel on the Big Island, on the first leg of what was to be an epic, round-the-world trip. We were having a perfectly wonderful time, when after breakfast one morning Satinder began having trouble breathing. She was also running a temperature. Concerned that she was suffering from an island fever of some sort, I called down to the front desk for a doctor. He showed up in about 15 minutes with a small, well-worn doctor's bag and quickly performed a brief exam before sharing his prognosis with me. He wasn't certain, but suspected that Satinder may have suffered a heart attack. In any event, he didn't think that the Big Island had the medical resources to do a thorough workup and recommended that we waste no time and take a private plane from the nearby airstrip to a hospital in Honolulu.

The pilot, Satinder, and I were the only ones on the plane. All I remember of the trip was that it was rocky and incredibly loud. It took almost an hour, and when we landed we took a cab directly to the hospital, uncertain of where or when our next stop would be or where we would spend the night.

After hours of poking, prodding, x-rays, and blood draws, our Honolulu doctors declared Satinder in fine health, the victim of a 'temporary episode' that they were unable to further diagnose. Still, just to be on the safe side, the doctors urged that she spend the night in the hospital. I joined her there, sleeping in a chair next to her bed.

The next morning she was breathing normally, had no fever, and looked – and said she felt – terrific, insisting she was ready to continue our trip. I was less certain than she about exposing her to the rigours of travel, so I arranged a few days of rest and further relaxation in Honolulu before we took off for Tokyo for another few days' stay, followed by a few more days in Hong Kong, then on to Delhi for three weeks.

If Satinder was ill in any way, she certainly wasn't showing it. Nor was there a scintilla of evidence suggesting another 'temporary episode' in the offing. On the contrary, she displayed nothing but eagerness and energy for the journey. The doctors in Honolulu were right, it seemed. She was completely fine.

It wasn't until we arrived and hunkered down in our apartment in London, our last scheduled stop on our round-the-world journey, that I noticed that her hand was shaking. It was over tea that I heard her cup rattle in its saucer. Noticing it as well, she quickly set cup and saucer down on the small table beside her.

'What was that?' I asked, searching her face for a sign of worry, or at least recognition. But it was not to come.

'What?' she countered. 'What was what?'

What followed for Satinder and me over the next many years was an odyssey in search of a cure for her illness that would take us thousands of miles across three continents to physicians, hospitals, yogis, clinics, healers, experts in the field, surgeons, fakirs, and charlatans. There were so many it was hard sometimes to discern which was even which.

That day in London when I first noticed the shaking, my impulse was to immediately take her to a hospital, or at least to a doctor. But Satinder would have nothing of it. As she put it, she didn't want to interrupt our vacation 'with a little nervous tremor.' So it wasn't until we arrived back in the States two weeks later and spoke to her internist in Palo Alto that we got the news that Satinder might be suffering from the early stages of Parkinson's disease.

I was, of course, seriously worried, but Satinder seemed unfazed, even sanguine. She demonstrated with another cup and saucer that the tremor was gone. And lo and behold, it seemed to be. Still, unconvinced that Satinder's earlier shaking cup was just another temporary episode, I booked passage for us back to London, to the Imperial College medical school where some of the best Parkinson's work was being performed. They confirmed the Parkinson's diagnosis and ran some further tests.

After that we were off to Stockholm, another hub for advanced Parkinson's research. There Satinder and I witnessed a still-experimental operation on the brain of a fully conscious Parkinson's sufferer, in an effort to significantly minimize the effects of the disease in its advanced stages. With her disease still in its early stages, Satinder would not be a likely candidate for the surgery, if and when it was further refined and became an approved procedure in the United States. Or so we were told by the Swedish researchers.

The most valuable single takeaway from our Scandinavian adventure was provided us by the director of the Swedish programme when he gave us the name of the very best person on Parkinson's in the world. The director scratched the name and address of this Parkinson's luminary on the back of one of his own business cards: Dr. J. William Langston; Parkinson's Institute and Clinical Center; Sunnyvale, California. Which is to say, we had travelled nearly 6,000 miles to learn that the world's top Parkinson's authority was in practice less than 15 miles from our home!

Back in California, Dr. Langston ran his own battery of tests, largely confirming what we already knew. By then we also had been informed that Parkinson's worked quickly in some cases but slowly in others, and that its effects and their seriousness varied widely. Typically an optimist, I simply assumed that Satinder's Parkinson's would be among the slower-to-develop cases. I was also heartened to learn that there were medications to help control the symptoms. But most of all, I took faith in Dr. Langston's prediction that a cure was only three to five years away. That was in 2000, and no cure is on the horizon even today. Those suffering from the disease are still being treated by the same medications. Nevertheless, whether the cure would come three or five years down the line, in 2000 Satinder and I were not about to just sit back and wait. Waiting was no more in her nature than it was in mine.

Our first major stop on the 'Cure Tour' was in India, where we stayed for nearly a month as Satinder spent her days working with a yogi known for his ability to help the ill and the infirm.

Returning from India, we focused our attention on a well-regarded Nepalese healer in Los Angeles whom Satinder discovered while researching non-Western medicine, a reputed expert in the treatment of Parkinson's symptoms. Sessions with this healer took place weekly, and I arranged for a driver to take her from our home directly to the healer's studio – 350 miles each way – to relieve her from having to navigate airports and cabs, as it was getting somewhat difficult for her to do so by herself, even with someone holding her arm. Also, in the early going, though her walking was not as smooth as it used to be, she absolutely refused a wheelchair.

Very therapeutic – or so it seemed – and fun for us both were our stays at the Hyatt Hotel in Carmel, a small coastal California town about an hour and a half south of Woodside, where we'd stroll through the picturesque village, or wander down the beach, or walk the town's quaint side streets and pathways, enjoying intimate lunches and dinners.

About fifteen years into Satinder's illness, she, along with Kiki, Raj, and I, decided it was no longer safe for her to drive. I know she hated giving it up – and, with it, her individual excursions into the world – but she acquiesced. Right about that time, too, she started using a walker. Life went on, though at a slower pace. I could tell that Satinder was still enjoying life – her times with me, with the kids (who were now in their late forties), and the grandkids. And Satinder never hungered for company. There were plenty of family get-togethers, house parties, and simple visits by friends.

Then, one Sunday afternoon in 2014, Satinder, still in bed, gasped, 'Help! I can't breathe!' I was dressing in the next room only a few feet away, but could barely hear her. I ran to the bed where I helped her stand, then sit, then stand again, trying to 'prime the pump,' to get her breathing comfortably again. But nothing seemed to work. Her breathing was becoming more and more laboured. I called an ambulance. Fifteen minutes later she was at Stanford Hospital in the ICU. By the time I'd parked the car, the doctors and the techs had her breathing stabilized. She looked a little lost, my precious Satinder, propped up on the starched white hospital bed. An oxygen mask helped her breathe, and a bank of wall-mounted monitors provided a constant readout of her signs, vital and otherwise.

In the hours that followed her admission, I counted eight doctors ministering to her, checking signs, comparing notes, writing on the clipboard at the foot of her bed. The collective assumption was that she had been ambushed by a side effect of Parkinson's. That, plus she had pneumonia. 'A temporary episode?' I ventured. The doctors concurred.

Three days into her stay at Stanford Hospital, one of the doctors suggested that Satinder would soon be 'out of the woods,' cured of her pneumonia, though she had yet to utter a word. It was then, as the doctor was trying to allay our concerns, that we noticed something odd – and, as it turned out, I wasn't the only one. Kiki and Raj had, as well, but were reluctant to suggest anything to contradict the prognosis of such a high-powered gathering of elite physicians. Meanwhile, it was obvious to us, at least, that, as animated as Satinder was with her right hand, she never moved her left. Rather, it lay limply on the sheets. I called this to the attention of one of the doctors who, after running some quick tests, turned to me to confirm the obvious: that sometime during Satinder's hospitalization, she had suffered a stroke.

And – as he failed to mention – that she had done so right under the noses of a battery of doctors in one of the country's finest teaching hospitals. Unbelievable, I thought. My initial feelings were of disbelief and denial. But they quickly devolved into anger.

'But you *could* have done something if you had caught it earlier,' I insisted. 'Or made it easier for her.' I was furious now. 'You could've *saved* her!' The doctors nodded, 'Perhaps so.'

The good news was – and at least there was *some* of that – that by day four of Satinder's hospital stay, she was talking as lucidly as someone could at eighty-four years of age, while fighting Parkinson's in a strange bed in a strange place, recovering from a stroke, and pumped full of drugs. But she was half paralysed and could no longer get along with only a walker. She had become seriously infirm almost overnight.

Earlier that afternoon, Raj, Kiki, and I sat in a hospital conference room together with three members of Satinder's 'team.' They had not asked before – an 'oversight,' they said – for a medical directive. Talk about doing things ass-backwards!

'How hard…' the most senior in age of the doctors asked, then began again: 'What measures do you want us to take to save her if it comes down to that? Even if, in the future, she has another episode and it is unlikely she'll regain consciousness. Or if…' He paused. I looked over to Raj, who was holding his mother's hand. Without giving it more than a moment's thought, I said, 'Every measure. I just want her to survive.'

And survive she did. Miraculously. And for over two years she lived comfortably with me in our home. Yes, she was largely bed- and wheelchair-ridden. But she was being seen to 24/7 by some of the most loving and giving professional caregivers available. Regular physical therapy sessions also made it easier for her to move and to try to regain her strength.

Meanwhile, for most of those two years she was completely *compos mentis,* only drifting off or not making sense every so often as she got older and her body started failing her more and more.

It was incredibly sad for all of us to see her so diminished, and at some point it became apparent that our collective hope to have her 'normal' again would never be realized.

It was around that time – when it appeared that we would truly lose her, and likely sooner than later – that I recalled a visit I had paid years earlier with a colleague of mine to the sickbed of a mutual friend of ours, Art Schawlow, another man of science, a Nobel Prize winner who was in a rest home in Palo Alto, with only weeks to live. The fact that he was in bad shape was underscored by his secretary, who cautioned us not to spend more than 15 minutes with him, and that any more might just take too much out of him.

She led us to the private room where Art was lying half asleep in a larger-than-standard hospital bed. Seeing us, he propped himself up and shook both our hands. His skin was incredibly dry but his handshake was firm. 'I'm sure she told you, "only 15 minutes," but I'm hoping it will be longer. I'm bored to tears here.' He sounded serious, so I promised that we would stay for at least a half hour.

Art was bartering for yet more minutes at the 30-minute juncture when I stopped him, myself feeling the press of time, not mine, but his: 'What do you worry about the most?' I asked. He'd lost his wife a few years back and I knew he had an intellectually challenged son. I spoke his name.

'He's being well taken care of,' Art said, 'but I worry about him, nonetheless. What happens when I'm not there?' He paused. 'Nothing for you to concern yourself with, in any event,' he pronounced, then paused. The next words he spoke were considerably softer. 'I *do* worry,' he began. 'I *do* worry about what happens after you die.' I looked to our mutual friend sitting across from me and he looked back to me. Clearly, he wasn't about to let me off easily. After all, I was the wise man, the one with the grey beard and turban, the one with the gravitas supposedly inherent in my advanced years.

So I told him what I had read and heard hundreds of times from the Guru Granth Sahib, the Sikh Holy Book, whose 1,430 pages, set to Indian classical music, basically says that life emerges like a spark from a fire, displaying its beauty for a time, only to fall back into the fire where that life becomes part of the universe.

I stopped, uncertain whether my recitation was providing Art the answer he was looking for or not, or whether he had even been paying attention. It was Art who broke the silence in the room by sitting up firmly and asking, 'So, what have you been up to, Narinder? Still inventing? More patents?'

I told him about the exciting new company I was putting together, the fibre-optic and laser pieces I had sculpted, the business meeting I was about to attend later that afternoon.

The more I rattled on, the more Art smiled, the more he looked like the familiar Art we used to know. 'Marvellous, marvellous,' he would interrupt occasionally. About the new company I hoped to launch, he asked that he be put on the family-and-friends list for stock. And the interesting thing to me was not so much his asking, but that he *meant* it, even though he must have known on some level that he only had weeks to live. Yet, he thought and talked about the future as if he would be playing an active role in it.

'Of course,' I said, 'I'll definitely be setting some stock aside for you.' He patted my hand as I held his other.

'And now, I'm getting just a little tired,' Art said, seeming to withdraw slightly from us, from the visit. My friend and I took the cue and after saying our goodbyes – all of us, I'm certain, aware that they might be our last – left the room.

'What do you think?' I asked my friend as we made our way down the corridor. 'Was he just fighting it? Denying it? The inevitability? Or was he only playing a game? Setting aside a darker reality for a lighter one?'

Art died, not long after.

Contemplating his passing today, I wonder, if at 90, I'm not playing the same game. Fielding phone calls and answering emails, writing a book, putting together yet another business, arranging for another exhibition, spending time with friends and family…?

One particularly bright note about the entire sad state of affairs surrounding Satinder's failing health was observing Kiki care for her mother. Watching her was a revelation! Satinder and I always knew that Kiki was an exceptional young woman and daughter, one who would do anything for those she loved. And when Satinder suffered, Kiki was there – overseeing the caregivers, conferring with the doctors, and spending hours every day with her mother: all this while working full-time as a lawyer and wife and mother to two children of her own. I have been truly blessed by the women closest to me in my life.

The one who loved me longest, Satinder, my beautiful, beloved wife, died on 25 June 2016.

Oh, Satinder...

Oh, Satinder! So much of what was *my* success – in business, in science, in the family, in collecting art, in life – was truly *ours*. I couldn't have done half the things I've done without you and your never-ending support. No one ever loved me as well as you. I know that. I always knew that.

And then there were the 'little' things. Don't think they went unnoticed, starting with the most basic. How in seemingly no time you transformed yourself from a passable cook into one of the finest anywhere. And not only making Indian dishes, but truly mastering a world cuisine.

And how, when I would ask the undoable, you would always manage to get it done. A dinner for twenty-five tomorrow evening in our home? No problem! A quick stopover to Sotheby's in London the day after tomorrow to preview a single piece of art and render your opinion? Sure. Your judgment always informed mine.

And how in 1984, when I became heavily involved in Sikh concerns with the Indian government and spent many a day and night in New York or Canada or somewhere else, you were always tending the home front. Day after day, you made the impossible possible, staying by my side.

And what a terrific mother you were: the care, the attention, the sheer love you had for Raj and Kiki both. It is no wonder that both of them miss you now almost as much as I do.

And then, of course, there were the *houses*. Not only the two on Greenways Drive, *our* houses, but the dozens of others in Palo Alto that you bought over time – your own highly successful business – after the kids left for college. Fixing them up, renting them out, managing the properties, and after a time selling them and moving on. The very office I'm in now. You convinced me to buy it when I was reluctant. Today it is worth fifty times what you suggested I offer… and I paid.

And then there was the house on College Avenue in Palo Alto that the owner simply *gave* you in her will. She loved the houses, she'd seen how well you treated neighbouring houses, and she knew you would take care of them, only selling if it became absolutely necessary. And we did take care of them, completely remodelling them, almost everything in the houses or on the property completely new. That was decades ago, and we still own them.

And probably most amazing – and a testament to your exceptional nature and generous spirit – in the twenty-six years you were in the real estate business, dealing with buyers and sellers and their agents and literally hundreds of tenants, you never once had a dispute. Not a single issue with anyone: remarkable.

Kiki, Satinder, and Raj at the Kapany home, Woodside, California.

Satinder Kaur Kapany, oil on canvas, Devender Singh, Kapany Collection.

Eulogy

Satinder loved chocolate.
She also loved *me*. More than anyone in the world, calling to me from our room as she neared the end of her life, 'Daddy, Daddy!' just to have me with her. Simply to comfort her as she felt the world slipping from her. It was a special treat for me, too, to have the time to spend, to be loved as well as I was by her.

Of course, besides chocolate, she loved Raj and Kiki, their spouses, and our grandchildren, Ari, Misha, Nikki, and Tara. Not long after Satinder passed, Raj said to me, 'Dad, I think mom's dying is getting harder for me to bear, not easier.' It's because she was a saint, no ordinary woman, the way she got along with everyone. In business and in all her personal dealings. And no matter what religion someone else might be.

It was Satinder's love for chocolate that had Kiki and me laughing again for the first time in what felt like a long time. The two of us were going through some bundles of Satinder's papers about a month after she died when an empty chocolate wrapper fluttered to the floor. It turned out to be the first of many, bundle after bundle, wrapper after wrapper. The two of us laughed so hard we cried.

An inherently jolly man, I was anything but jovial in the days and months following Satinder's death, in August of 2016. At my age – ninety-three as of this writing – I'd experienced the passing of many of my friends, relatives, and colleagues and was always able to keep their dying in perspective. We're born, we live, we die. Each and every one of us.

Foolishly, I believed it would be that way with Satinder's passing. But it wasn't. It *isn't.* I walk through the house and witness her decorator's touch, her impeccable taste. It is everywhere in evidence. I open her closets, see her clothes, get a whiff of her essence. I walk through the garden, marvel at the incredible blooming plants, at the mature fruit trees… furious all over again about that damned fertilizer, wanting nothing more than for her to be present in our lives once again.

Not long after Satinder's passing, I had lunch with a close friend, Dick Pantell, the same friend who'd accompanied me to Art Schawlow's hospital bedside years ago, a retired Stanford engineering professor – which is to say, another man of science. His own wife had also died of cancer a few years back, but on the day of our lunch he confessed that for the longest time after his wife's passing he'd carry on conversations with her that were as real to him as if she were actually in the room. 'Do you also do that, Narinder?' he asked. 'No,' I said, but then added, 'though, after giving it some thought, I'm not ruling it out for the future.'

Well, the future has arrived. It's both come and gone, in fact, and I've had only a very few discussions with Satinder who – sadly for me, for Raj and Kiki, and for the grandchildren – is also gone, and *undeniably* so.

Still, I take solace in the fact that her spark shone so very brightly for her time with us. And then it inevitably faded to mingle with all the other spirits in the universe, past, present, and future.

To the years gone by, I bid a fond farewell. For those yet to come, I welcome them, I embrace them.

Wildlife photographed on a family safari in South Africa,
Picture courtesy: Misha and Michael.

FAMILY PETS

Narinder and Satinder with their pets.

To To II.

Jo Jo.

To To.

Left: Budhoo the pet monkey grooming Kali the cat at Dehradun, India.
Right: Raj leading Myneyn the Arabian horse.

Left: Myneyn, enjoying a gallop. *Right:* Atilla, catching her breath after a run.

In Closing

To the years gone by, I bid a fond farewell.

For those yet to come, I welcome them, I embrace them.

And to *you*, I offer these closing thoughts, remembering back, as I do, to the early 1930s when I bathed with my uncle in Moga at the well on our property. Back then the water level was less than 10 feet from the surface, the water cool and clear enough to drink. Today, however, you have to drill down many hundreds of feet to reach water. And when you do, it is often polluted and undrinkable.

When I left my home country for London in 1951, India's population was 350 million. Today it is more than 1.3 billion. Cities have become overpopulated, while pollution, as much as six times worse today than in the 1950s, is out of control, with everyone affected – from the young entrepreneurs in Western garb thriving in India's software industry to the ragged children living on the streets.

Of course, a burgeoning population with rampant pollution is hardly exclusive to India. These issues and a host of others even more insidious are bedevilling countries worldwide. A superheated atmosphere has led to the melting of huge icebergs, as many countries lose their land to a rising tide. This has already happened in some regions, and is likely to accelerate in the near future.

Wildfires in California and Australia have devastated hundreds of thousands of acres and more, in part the result of drastic temperature change and droughts, with other climatic changes bringing about a variety of disasters.

Add to these volcanic eruptions, earthquakes, tsunamis, floods, hurricanes, tornadoes, global pandemics and other natural occurrences and it is clear that today's youth – that's *you!* – have more than got their work cut out for them. While the new science they bring to the task may be able to ameliorate some of these calamities, it's unclear what will be the consequences of their timing and their cost.

Even more challenging in meeting these human-made and natural issues will be to do so humanistically, avoiding the recent international political movements that favour 'strict nationalism' and racial and religious profiling. Together, we must work toward absolute equality for all, promoting love and charity.

This is not *one* way for moving forward; it is the *only* way.

Narinder Kapany, March 2020